Robin Nittka

Elliptic and parabolic Robin problems on Lipschitz domains

Robin Nittka

Elliptic and parabolic Robin problems on Lipschitz domains

Hölder continuity of solutions of elliptic problems and generation of nonlinear semigroups on the space of continuous functions

Südwestdeutscher Verlag für Hochschulschriften

Imprint

Publisher:
Südwestdeutscher Verlag für Hochschulschriften
is a trademark of
Dodo Books Indian Ocean Ltd., member of the OmniScriptum S.R.L Publishing group
str. A.Russo 15, of. 61, Chisinau-2068, Republic of Moldova Europe
Printed at: see last page
ISBN: 978-3-8381-1898-7

Zugl. / Approved by: Ulm, Universität Ulm, Dissertation, 2010

Preface

As the title already suggests, in this thesis we are concerned with elliptic and parabolic problems. Roughly speaking, our investigations of the solutions can be separated into four categories: existence, uniqueness, regularity, and continuous dependence on the data. The emphasis is on the regularity for elliptic and parabolic equations. But it turns out that in order to study regularity one has also to understand the other three topics. We start with a short description of our motivation on the basis of a standard example.

One of the model equations in the theory of partial differential equations is the Neumann problem

$$\begin{cases} \lambda u - \Delta u = f & \text{on } \Omega, \\ \dfrac{\partial u}{\partial \nu} = g & \text{on } \partial\Omega \end{cases} \tag{1}$$

with $\lambda > 0$ and Ω an open set in $\mathbb{R}^N$. A basic result is that if Ω, f, and g are of class C^∞, then there exists a classical solution $u \in \mathrm{C}^\infty(\overline{\Omega})$ of (1).

In many important situations one does not have such a high regularity. For example for numerical applications it is very natural to consider (1) for a domain Ω that has corners, maybe a triangle in $\mathbb{R}^2$. Passing to a weaker notion of solution, the celebrated Lax-Milgram theorem tells us that there exists a unique weak solution u of (1) even if Ω is merely a Lipschitz domain, where we only have to require $f \in L^2(\Omega)$ and $g \in L^2(\partial\Omega)$. Here, a weak solution is a priori only once weakly differentiable, i.e., $u \in W^{1,2}(\Omega)$.

It is now a natural question whether this weak solution has higher regularity. If Ω is smooth enough, then the unique weak solution u of

$$\begin{cases} \lambda u - \Delta u = f & \text{on } \Omega, \\ \dfrac{\partial u}{\partial \nu} = 0 & \text{on } \partial\Omega \end{cases} \tag{2}$$

is twice weakly differentiable, and more precisely $u \in W^{2,p}(\Omega)$ if $f \in L^p(\Omega)$. Thus if $f \in L^{q/2}(\Omega)$ for some $q > N$, then the unique solution of (2) is continuous up to the boundary of Ω, i.e., $u \in \mathrm{C}(\overline{\Omega})$, by the Sobolev embedding theorems, and in fact even Hölder continuous up to the boundary. On the other hand, for general Lipschitz domains the solution will typically not even be in $W^{1,p}(\Omega)$ for large p even if f is very smooth. We can, however, still hope that u is continuous up to the boundary if $f \in L^{q/2}(\Omega)$ for some $q > N$.

In fact, the desire to prove that the weak solution u of (2) is continuous up to the boundary if Ω is a bounded Lipschitz domain and $f \in L^{q/2}(\Omega)$, $q > N$, was the starting point for this thesis. The basic approach to attack this problem is to extend u by reflection along the graph of $\partial\Omega$ to a function $\tilde{u}$ defined in a neighborhood Ω' of $\overline{\Omega}$. We can show that $\tilde{u}$ solves again a uniformly elliptic differential equation on Ω'. But since the derivative of a Lipschitz function is in general not continuous, the coefficients of this new equation are merely bounded. Still, the celebrated result independently due to De Giorgi and Nash allows us to deduce that $\tilde{u}$ is continuous on Ω'. Hence u is continuous on $\overline{\Omega}$.

It turns out that this strategy still works for the inhomogeneous equation (1). Moreover, the proof goes through for a much larger class of differential equations, including degenerate and singular quasilinear elliptic equations. In addition, it is possible to obtain the analogous result for Robin boundary conditions via reduction to the Neumann case. Finally, we can apply these elliptic results to show that the corresponding parabolic problems are well-posed in the space $\mathrm{C}(\overline{\Omega})$, i.e., we have continuous solutions for continuous initial data, or, in other words, the elliptic operator generates a strongly continuous semigroup on $\mathrm{C}(\overline{\Omega})$.

We now discuss the general setting. In this thesis we consider quasilinear, elliptic, second order partial differential equations in divergence form with inhomogeneous Robin boundary conditions on a bounded Lipschitz domain $\Omega \subset \mathbb{R}^N$, i.e., equations that essentially look like

$$\begin{cases} -\operatorname{div} A(x,u,\nabla u) = f & \text{on } \Omega, \\ A(x,u,\nabla u)\cdot\nu + h(u) = g & \text{on } \partial\Omega, \end{cases} \tag{3}$$

where ν denotes the outer unit normal of Ω. The model problems we have in mind arise for

$$A(x,u,\nabla u) = (1+\arctan|\nabla u|^2)\,\nabla u \tag{4}$$

and

$$A(x,u,\nabla u) = |\nabla u|^{m-2}\nabla u \tag{5}$$

with $m \in (1,\infty)$. For the latter, problem (3) is called the m-Laplace equation. Equations of this type appear in many mathematical models of physical processes, e.g., nonlinear diffusion and filtration [Phi61], deformation plasticity [AC84], and viscoelastic materials [LT90]. Also the special case of linear equations is included, and frequently we pay special attention to it. Our main results for these problems are summarized in Examples 3.2.7, 3.2.9, 3.2.10, and 3.2.16.

Several authors have shown under varying assumptions on the coefficients that every solution of (3) is locally Hölder continuous in the interior of Ω, see for example [LU68, Ser64, Tru67]. Without exaggeration, it can be said that it is well understood how interior regularity can be obtained for equations like (3).

Regarding regularity at the boundary, classically the domains are assumed to be smooth. More recently, however, the class of Lipschitz domains has attracted much interest. As examples we refer to the articles by Kenig and Rule [KR09], Mitrea and Monniaux [MM09], Mitrea and Taylor [MT09], Agranovich [Agr08], Shen [She08], Haller-Dintelmann and Rehberg [HDR09], and Wood [Woo07], to name only a few of the newest contributions.

Among all possible boundary conditions, Dirichlet boundary conditions are the most popular, meaning that one prescribes values for u on the boundary of Ω. Regularity of solutions of Dirichlet problems lies at the very heart of potential theory, and sharp conditions are known under which the solution is continuous up to the boundary of Ω. For the linear case, this is the celebrated Wiener criterion. Quasilinear generalizations have been found and studied by Maz'ya [Maz70], Gariepy and Ziemer [GZ77], and Kilpeläinen and Malý [KM94]. In fact, much is known about the regularity of the solution and its derivatives even if the right hand side is very rough, see for example recent articles by Mingione [Min07, Min10] and Duzaar and Mingione [DM09].

For Robin boundary conditions, on the other hand, and even for the special case of Neumann boundary conditions, i.e., if $h(u) = 0$ in (3), the situation is not as well understood. There are, however, results due to Lieberman [Lie83, Lie92] if the domain is smooth except for a small set. One of the main goals of this thesis is to establish regularity up to the boundary also for these boundary conditions if Ω is a Lipschitz domain. More precisely, we want to show that every solution is Hölder continuous up to the boundary of Ω. This means that u allows for a continuous extension to $\overline{\Omega}$ which is Hölder continuous for the same exponent and the same Hölder constant.

For Neumann boundary conditions, we use the elegant method that we had already described for the Laplace operator to deduce regularity at the boundary from interior regularity results. In fact, if u is a solution of (3), then we extend u to a function $\tilde{u}$ on a larger domain Ω' containing $\overline{\Omega}$ by reflection along the boundary of Ω. If $g = 0$, then $\tilde{u}$ solves an equation of the same type as (3) on Ω'. Thus we can deduce that $\tilde{u}$ is locally Hölder continuous in the interior of Ω' and hence in particular on Ω, where it coincides with u.

Let us compare this trick with the standard approach to regularity at the boundary: for smooth boundary, a typical technique is to locally transform the problem into a differential equation on a half-space by a smooth, nonlinear transformation in the spatial variables. Regularity on a half-space can often be shown by direct calculations. Alternatively, one may extend the function from the half-space to the whole space by reflection and apply the same arguments as above. In this sense, our idea is somewhat similar to the usual one, but we merged the two steps into a single one. Moreover, due to the lack of smoothness of the boundary, we cannot apply the usual spatial transformation, which would correspond to reflection along the outer normal of Ω, but have to reflect along the graph.

There are still some details that require additional care. Firstly, we run into trouble because the structure of the equation is not maintained under reflection if $g \neq 0$. However, this is a minor issue. In fact, the extension $\tilde{u}$ still solves an elliptic problem, but with a distributional right hand side, and interior regularity results are known also in this case.

Secondly, a major difficulty arises if we want to allow non-trivial functions h. The strategy the author found to overcome this issue is as follows: fix a solution u of (3) and consider

$\tilde{g} := g - h(u)$ as fixed. Then (3) takes the form

$$\begin{cases} -\operatorname{div} A(x, u, \nabla u) = f & \text{on } \Omega, \\ A(x, u, \nabla u) \cdot \nu = \tilde{g} & \text{on } \partial\Omega, \end{cases}$$

i.e., (3) is an inhomogeneous Neumann problem. We can hope that the result for the Neumann case applies to this situation. In typical situations, however, it is not obvious that $\tilde{g}$ is regular enough to apply the result for Neumann equations. This is where the technicalities start.

In order to show that $h(u)$ is well behaved, we use nonlinear interpolation and bootstrapping. More precisely, we assume that the Neumann problem has a unique solution for every right hand side, its resolvent is Hölder continuous, and that for sufficiently regular right hand side the solution is Hölder continuous. These assumptions allow us to apply a nonlinear interpolation theorem in order to deduce that the resolvent is regularizing on a scale of L^p-spaces. Now a bootstrap procedure gives the desired regularity for $h(u)$ and hence Hölder continuity of u. We also exhibit large classes of equations that satisfy these assumptions. In particular, the aforementioned model cases are covered.

Finally, we apply the elliptic theory to parabolic problems. In fact, having very good elliptic results at hand, it is not difficult to show that a parabolic problem like the heat equation with Neumann boundary conditions

$$\begin{cases} \dot{u}(t) = \Delta u(t) & \text{on } \Omega \text{ for } t > 0, \\ \dfrac{\partial u(t)}{\partial \nu} = 0 & \text{on } \partial\Omega \text{ for } t > 0, \\ u(0) = u_0 & \text{on } \overline{\Omega} \end{cases}$$

is well-posed on $C(\overline{\Omega})$, provided one knows the appropriate theorems of semigroup theory. Here, by well-posedness we mean that for every continuous function u_0 there exists a unique solution in a weak sense and that this solution, considered as a function on the parabolic cylinder $[0, \infty) \times \overline{\Omega}$, is continuous. We prove this regularity for general linear equations with Robin or Wentzell-Robin boundary conditions and also for the two quasilinear parabolic equations with Robin boundary conditions arising from our model examples (4) and (5).

We conclude the introduction with a more precise outline of the results, including citations to related research articles and books.

Chapter 1 contains the definitions and fundamental results that we apply in the main body of this manuscript. More precisely, Section 1.1 consists of several estimates for real numbers, or rather for vectors in $\mathbb{R}^N$. These are well known to people working with the m-Laplace equation. In Section 1.2 we introduce the Sobolev spaces, in which weak solutions are to be found, and spaces of Hölder continuous functions. In this context, we recall several rules for the calculus of Lipschitz continuous functions such as the change of variables formula and the chain rule. Section 1.3 addresses the geometric notion of a Lipschitz domain, which comprises the class of polygons that is particularly important in applications. We quote some analytic facts for later use, most prominently the divergence theorem, the Sobolev embedding theorems and the trace theorems. Section 1.4 concludes the elementary part of the preliminaries with a very short summary of selected theorems from functional analysis that are not entirely renowned, which is included for easier reference. For an introduction to the language of functional analysis, which is used freely, the reader is referred to the books by Werner [Wer00] and Lax [Lax02].

Starting with Section 1.5, which contains existence theorems for nonlinear equations, the material becomes more sophisticated. There are only few general ideas to solve quasilinear equations. A particularly successful one is the approach via monotonicity conditions, which we use here. The theory of monotone operators goes back to Golomb [Gol35], Kačurovski [Kač60], and Zarantonello [Zar60], and it has been developed much further by many authors, among them Brézis, Minty, Browder, Rockafellar, Leray, and Lions. It generalizes to the larger class of pseudo-monotone operators introduced by Brézis [Bré68], which is the type of operator that we consider here because this class is more stable under perturbations. We provide the basic definitions of the theory and the results that are important for our applications, the main source being Showalter's book [Sho97]. Detailed information about monotone operators can also be found in Brézis' book [Bré73].

Section 1.6 is divided into two parts, both concerning interpolation theory. For general information about interpolation theory we refer to Triebel's book [Tri95]. The first part of Section 1.6 is loosely related to interpolation theory, but rather a direct estimate bounding a Hölder continuous function in terms of its Hölder norm and an L^p-norm. This result due to Lê [Lê07] is interesting from the point of view of interpolation theory because it tells us that $L^\infty(\Omega)$ is of class $J(\theta, L^q(\Omega), \mathrm{C}^{0,\alpha}(\Omega))$, see [Tri95, §1.10.1]. The purpose for which we use it is the following: if u and $\tilde{u}$ are solutions of a quasilinear equation of the form (3) for right hand sides (f, g) and $(\tilde{f}, \tilde{g})$, and if we know that we can bound u and $\tilde{u}$ in $\mathrm{C}^{0,\alpha}$ in terms of the norms of f, $\tilde{f}$, g, and $\tilde{g}$, and if we know in addition that we can bound $u - \tilde{u}$ in $L^m(\Omega)$ by $\|f - \tilde{f}\|^\beta + \|g - \tilde{g}\|^\beta$ for some $\beta \in (0,1)$, then the above estimate means that the resolvent of the equation is locally Hölder continuous. In fact, this is precisely the type of argument for which Lê used his estimate.

The second part of Section 1.6 is at the heart of interpolation theory. Essentially, we give sufficient conditions under which a nonlinear operator that maps a Banach space A_1 into a Banach space B_1 and a subspace A_0 of A_1 into a subspace B_0 of B_1 does in fact map real interpolation spaces between A_0 and A_1 into real interpolation spaces between B_0 and B_1. For linear operators, this belongs to the very essence of interpolation theory. For nonlinear operators, the situation is much more difficult and less understood. For the complex interpolation functor, the main results are due to Bergh [Ber84], Cwikel [Cwi87], and Böhm [Böh91, Böh93]. A short survey on the much richer nonlinear real interpolation theory can be found in an article by Berezhnoj and Zabrejko [BZ88]. Here, we follow the ideas due to Peetre [Pee70] and Tartar [Tar72]. In fact, all necessary ideas for our abstract nonlinear interpolation result are already contained in [Tar72], but the conclusion is missing. Thus, and because the application to products of Lebesgue spaces needs an additional argument, we include proofs in that section. For these, some auxiliary information on interpolation spaces is provided as well.

The material about one-parameter operator semigroups that is needed in Chapter 4 is included in Sections 1.7 and 1.8, the former concerning the linear case, whereas the latter treats nonlinear semigroups.

We conclude the second chapter with an introduction to quasilinear equations in divergence form in Section 1.9. In fact, the first part of that section is concerned with the definitions of interior weak solutions and solutions of boundary value problems. We follow the usual approach by defining a weak solution in a distributional sense, i.e., we require that after integrating against a smooth function and formally integrating by parts, we have equality of both sides of the equation. The second part collects results about interior regularity of solutions. The linear, symmetric case amounts to the celebrated result due to De Giorgi [DG57] and Nash [Nas58], later reproved by Moser [Mos60], see also the recent survey on regularity results by Mingione [Min06] for an exposition of De Giorgi's method and some more recent developments. The generalizations of their techniques to larger classes of problems are way too plentiful to mention. But for our purposes in the quasilinear case, it suffices to refer to the results by Ladyzhenskaya and Ural'tseva [LU68], Serrin [Ser64], and Lieberman [Lie91] stating that every weak solution of a quasilinear problem whose coefficients satisfy very natural growth conditions is locally Hölder continuous. The special feature of these particular references is that the coefficients are allowed to be unbounded with respect to the space variable. For our applications, it is not necessary to allow unbounded coefficients, but since this comes at no extra cost, we include it here.

Chapter 2 contains the results for Neumann boundary conditions. More precisely, we exhibit several classes of Neumann problems, which include our prime examples, for which we have well-posedness and regularity. As for the regularity, in Section 2.1 we apply the reflection argument that has already been mentioned to obtain Hölder continuity up the boundary of Ω. It is easy to find an equation that is solved by the reflection of u along the boundary of Ω once we are in local coordinates. The main part of this section consists in checking that the coefficients of the new equation indeed are regular enough for the results about interior regularity to be applicable. It is worth mentioning that for this approach to work we need a very general version of the interior regularity result. In fact, it turns out that the new coefficients are discontinuous at $\partial\Omega$ and that the new right hand side is not a function, but merely a distribution, even if the original coefficients and the original right hand side are very smooth. Still, the assumptions under which we obtain Hölder regularity at the boundary are the same as the classical assumptions for interior regularity, which is clearly optimal for this approach. Hölder regularity for the inhomogeneous Neumann problem has of course already been proved by the usual techniques, see for example [Lie83]. However, our approach is much simpler. Surprisingly, it seems that nobody has exploited this idea to this extent before, even though the reflection is natural and several authors used it to some degree [Tol83, Tro87, Nit09a] or at least mentioned the possibility [BH91].

The other sections of Chapter 2 serve the purpose of providing large classes of examples for the theory of Chapter 3. In fact, in Chapter 3 we will assume that for every right hand side there exists a solution and that the solution depends on the right hand side in a Hölder continuous way. As for existence, in Section 2.2 we use the Lax-Milgram lemma to show that in the linear case we always have a solution. For the quasilinear case we use the theory of monotone operators. In both cases, an estimate for the norm of the solution in terms of the right hand side is automatic, and this will be used later.

In Section 2.3 we ask whether the solution depends on the right hand side in a Hölder continuous way. For linear equations, this is of course automatic, and we make this explicit in Subsection 2.3.1. For quasilinear equations, on the other hand, we obtain only partial results. Surprisingly, also in the literature not much can be found on this topic. Regarding an estimate for the difference of the gradients, confer the articles of Mingione [Min07], Kristensen and Mingione [KM08], and Duzaar, Kristensen, and Mingione [DKM07]. Biegert [Bie10] has a preprint where he proves locally Hölder continuous dependence for the m-Laplace equation. However, his proof works not for the whole range $m \in (1, \infty)$, but only if m is not too close to 1. This manuscript borrows some of his ideas, which can also be found in an article by Cheng [Che98]. The simple idea due to Lê [Lê07] which was mentioned before is applied in Corollary 2.3.19 in order to obtain locally Hölder continuous dependence. However, the nonlinear interpolation that we want to apply needs more precise information about the exponents that is not provided by Lê's method. Still, it is a remarkable fact that in all situations considered here we have locally Hölder continuous dependence for some exponent, and this fact is important also for the study of the parabolic problems.

In Subsection 2.3.2, we reduce certain very regular equations to the linear case by differentiation. More precisely, whenever u and v solve a quasilinear equation for different right hand sides and the coefficients of the equation are differentiable, then their difference solves a linear equation, where the new right hand side is the difference of the original right hand sides and the new coefficients are the derivatives of the original coefficients, but depend on u and v and their derivatives. If we want to benefit from this representation, we need control over the new coefficients. So either the derivatives of the coefficients have to be suitably bounded like in our model case

$$A(x, u, \nabla u) = (1 + \arctan |\nabla u|^2)\nabla u$$

or we need good estimates for the functions and their derivatives. Both alternatives are pursued in Subsection 2.3.2.

Finally, Subsection 2.3.3 features an entirely different approach to continuous dependence on the right hand side. If the coefficients are uniformly monotonic in a sense that will be made precise, it is trivial consequence that the resolvent is Hölder continuous. In fact, this argument is standard for the m-Laplace operator for $m \geq 2$, see for example the article by Biegert and Warma [BW09]. However, we never have this uniform monotonicity for $m < 2$ and therefore we need a somewhat different idea in this case. Still, under a natural assumption that is satisfied for the m-Laplace operator, a similar, but technically more difficult strategy works and gives even Lipschitz continuous dependence. However, this result is still worse than the Hölder continuous dependence for $m \geq 2$ regarding its relation to the growth behavior of the solution in dependence of the right hand side. This statement will become clear in Section 3.2 when we apply the nonlinear interpolation result. The idea for the proof in the case $m < 2$ is strongly motivated by Biegert's calculations [Bie10].

In Chapter 3, in principle we repeat the program of Chapter 2 for Robin boundary conditions. The focus is strictly on the regularity since on the one hand this was the main motivation from the beginning, and on the other hand the arguments for existence and continuous dependence are very similar to those for the Neumann problem. In fact, Section 3.1 contains only a short summary of what changes in the existence proof in the monotone and the linear case when we consider Robin boundary conditions, and the continuous dependence is entirely omitted.

The interesting part is the Hölder continuity of solutions, treated in Section 3.2. Our results seem to be new, at least in the nonlinear case. Weaker version of this result in the linear case can be found in [BH91, War06, Dan09, AR97], whereas a stronger result is due to Griepentrog and Recke [GR01]. In Section 3.2 we attack the problem on a rather abstract level in comparison to Chapter 2. In fact, we assume a certain well-posedness and regularity behavior of the corresponding Neumann problem in order to deduce Hölder regularity of the solutions of the Robin problem. More precisely, we show that the Robin term can be regarded as a regular inhomogeneity of the Neumann problem. For this, we have to prove that the solution of the Robin problem has a sufficiently regular trace, which we achieve by a bootstrapping procedure based on nonlinear interpolation. For $m \geq 2$, the result is satisfactory and the exponents that appear in the calculations are rather natural. Unfortunately, for $m < 2$ the exponents turn out to be quite a mess. Moreover, the strategy fails if m is too small compared to the dimension of the space. Still, the admissible range for m is larger than it is usual in these situations, compare for example [Bie10] and see also Remark 3.2.15. But the complicated expression for the lower bound of admissible m can be regarded as an indication that there is still room for improvements. For our prime example, the m-Laplace operator with $m < 2$, we could circumvent this whole problem by quoting an article by Daners and Drábek [DD09] who prove boundedness of u by Moser's iteration method, see Remark 3.2.17.

Finally, in Chapter 4 we discuss Robin and Wentzell-Robin parabolic equations as an application of the elliptic regularity result. These problems have been extensively studied, in particular from a semigroup perspective, and the articles [AtE97, Dan00b, AW03, AMPR03, CFG+08] constitute merely a small assortment of research articles on that topic. We use semigroup theory to show that linear parabolic problems with Robin or Wentzell-Robin boundary conditions are well-posed in the space $C(\overline{\Omega})$ if Ω is a bounded Lipschitz domain. By this we mean that for continuous initial data the solution is continuous in time and space. Weaker versions of this result can be found in various articles [BH91, FT95, FGGR02, Eng03, War06], but for arbitrary elliptic operators this seems to be due to the author, see also [Nit09b]. Our strategy is similar to Warma's [War06] who considered the Laplace operator only.

We start with the parabolic Robin problem

$$\begin{cases} \dot{u}(t,x) = -Lu(t,x), & t>0,\ x\in\Omega, \\ \frac{\partial u}{\partial\nu_L}(t,z)+\beta u(t,z)=0, & t\geq 0,\ z\in\partial\Omega, \\ u(0,x)=u_0(x), & x\in\Omega \end{cases}$$

in Section 4.1. Here, L denotes a second order, linear, divergence form elliptic operator with bounded coefficients. First, we need to know that the corresponding semigroup is well-defined and bounded on $L^\infty(\Omega)$. In most cases, this is easily shown by the Beurling-Deny criterion, compare [AtE97], but for arbitrary coefficients this is a delicate matter. Fortunately, a result by Daners [Dan00a] tells us that the semigroup satisfies Gaussian estimates, which easily implies boundedness in $L^\infty(\Omega)$. We show that the continuous functions are invariant under the resolvent of the elliptic operator, which is an easy consequence of the results in Chapter 3, and that the part of the operator in the space of continuous functions is densely defined, which is also a consequence of the results in Chapter 3, but needs a rather careful analysis and the full power of the linear results. More precisely, we have to admit distributions as right hand sides.

In Section 4.2, we study the parabolic Wentzell-Robin problem

$$\begin{cases} \dot{u}(t,x) = -Lu(t,x), & t>0, x\in\Omega, \\ -Lu(t,z)+\gamma\frac{\partial u(t,z)}{\partial\nu_L}+\beta u(t,z)=0, & t>0, z\in\partial\Omega, \\ u(0,x)=u_0(x), & x\in\Omega. \end{cases}$$

We follow the strategy due to Arendt, Metafune, Pallara, and Romanelli [AMPR03] and attack this problem by form methods. Then the same strategy as for the Robin problem works with only minor changes, apart from the fact that we do not have Gaussian estimates. As a substitute, we use the Beurling-Deny criteria to obtain boundedness of the semigroup on L^∞, which works in fact in more cases than for Robin boundary conditions. However, for this to work we have to introduce the unsatisfactory condition that the first order coefficients of the operator are Lipschitz continuous. This is of course no severe restriction, but it is nevertheless surprising, in particular because we need no regularity assumptions on the second order coefficients.

We conclude Section 4.2 by the observation that quasi-contractiveness of the Wentzell-Robin semigroup in L^p, $p\in[1,\infty]$, can by a witty trick be obtained from the corresponding property of the Robin semigroup. This seems to be new, and it extends a result due to Favini, Goldstein, Goldstein, and Romanelli [FGGR08].

In Section 4.3 we deal with quasilinear parabolic problems. If the domain, the coefficients, and the initial data are smooth, it is known that the solution of such a parabolic problem is jointly Hölder continuous in its time and space variables, and one has control over the gradient, see [DiB93]. If the boundary is only Lipschitz regular and the initial function is only continuous, we still expect the solution to be jointly continuous, and in fact this can be proved with the help of the elliptic results. To keep things reasonably simple, we examine only the parabolic problems arising from the two model operators (4) and (5) with Robin boundary conditions. The same strategy as for the linear case allows us to prove that the problem is well-posed in the space $C(\overline{\Omega})$. For Robin and even for Neumann problems on Lipschitz domains this seems to be new.

I wish to conclude this introduction by expressing my sincere gratitude to everybody who supported me during my studies and the work on this manuscript. Of course it is not possible to mention every single person who deserves it, so I restrict myself to the ones that are most closely related to the work on this thesis.

First of all, of course, I am deeply indebted to my advisor Prof. Wolfgang Arendt who has been accompanying my mathematical development for many years. I owe to him many insights, regarding both mathematics and mathematical style. I am also grateful to Prof. Werner Kratz for his kind readiness to be the second referee for this thesis.

I thank all the members and former members of the institute, with whom I had interesting discussions and generally a good time, but in particular Manfred Sauter who escorted me during the largest part of my university life. And I should not forget to mention John W. Neuberger who stimulated some research and made it possible for me to spend quality time collaborating with him.

Dr. Petra Kaiser deserves my most cordially thanks for her patience and support during the last few years. Also my parents have my gratitude for their support and interest in what I was doing.

Lastly, I thank the graduate school MATHEMATICAL ANALYSIS OF EVOLUTION, INFORMATION AND COMPLEXITY for the financial support during a large part of the work on this thesis and for providing a platform for interdisciplinary exchange.

Contents

1 Preliminaries and notations

In this chapter we collect several tools that will be needed later. Most of the material of the first few sections is standard. It is included for easier reference, self-containedness, and to establish a common language. The common link between all the results of this chapter is that they are preparatory for the main results, which are contained in the later chapters.

In Sections 1.1 to 1.4, the basic analytic notions are summarized. Since the reader is assumed to be familiar with most of this content, the theorems are used in the sequel without further reference. This collection of results mainly serves the purpose to define what we understand here by several terms that might be used ambiguously in the literature. Sections 1.5 to 1.9 contain more specialized results that might not be common knowledge. Here, sketches of the proofs are included if this was possible with reasonable effort.

Throughout, all functions will be real-valued. The variable $N \in \mathbb{N}$ will be the dimension of the underlying Euclidean space. Except for some examples that for simplicity will be set up in one dimension, in the whole manuscript we will assume that $N \geq 2$. On the one hand, this is because the situation $N = 1$ is trivial for the main theorems. On the other hand, some expressions in the statements of the theorems would not be defined for $N = 1$, so we would have to distinguish that case from the others. To a certain degree, the same is true for $N = 2$. However, since this case is more important than $N = 1$, it will always be included.

1.1 Estimates

In this section we present some technical estimates. The first three are well known. They are included here nonetheless since usually less general versions are associated with the respective names.

Lemma 1.1.1 (Young's inequality, general version). *Let $N \in \mathbb{N}$, let $p_i \in [1, \infty)$, $i = 1, \ldots, N$, be such that $\sum_{i=1}^{N} \frac{1}{p_i} = 1$, and let $c_i > 0$ be such that $\prod_{i=1}^{N} c_i = 1$. Then*

$$\prod_{i=1}^{N} a_i \leq \sum_{i=1}^{N} \frac{c_i^{p_i}}{p_i} a_i^{p_i}$$

for all non-negative real numbers a_i, $i = 1, \ldots, N$.

Theorem 1.1.2 (Hölder's inequality, general version). *Let (Ω, μ) be a measure space, $N \in \mathbb{N}$, $p_i \in [1, \infty]$, $i = 1, \dots, N$, and $\sum_{i=1}^N \frac{1}{p_i} = \frac{1}{r}$ with $r \geq 1$. If $f_i \in L^{p_i}(\Omega)$, $i = 1, \dots, N$, then $\prod_{i=1}^N f_i \in L^r(\Omega)$ and*

$$\left\| \prod_{i=1}^N f_i \right\|_{L^r(\Omega)} \leq \prod_{i=1}^N \|f_i\|_{L^{p_i}(\Omega)}.$$

In the case when $N = 2$ and $r = 1$, we have $\frac{1}{p_1} + \frac{1}{p_2} = 1$ in Theorem 1.1.2. The unique number p' in $[1, \infty]$ such that $\frac{1}{p} + \frac{1}{p'} = 1$ is called the *conjugate exponent* of p.

Lemma 1.1.3. *For all $p \in (0, \infty)$ and $N \in \mathbb{N}$, there exist positive constants α and β such that*

$$\alpha \sum_{j=1}^N a_j^p \leq \left(\sum_{j=1}^N a_j \right)^p \leq \beta \sum_{j=1}^N a_j^p \tag{1.1.1}$$

for all $a_j \geq 0$, $j = 1, \dots, N$.

Proof. For $a_j \geq 0$ and $p > 0$ we have

$$\max_{j=1,\dots,N} a_j \leq \left(\sum_{j=1}^N a_j^p \right)^{1/p} \leq N^{1/p} \max_{j=1,\dots,N} a_j.$$

Thus the expression $A_p := \left(\sum_{j=1}^N a_j^p \right)^{1/p}$ is comparable with $\max_{j=1,\dots,N} a_j$ for every $p > 0$, i.e., each one can be estimated by the other with a constant that does not depend on the a_j. Hence also A_p and A_1 are comparable, which implies (1.1.1). □

The last two estimates of this section are sophisticated. Both are often used in the context of degenerate quasi-linear differential equations, but frequently without proof. The first one gives the more precise information where both apply, but it is valid only for $m \geq 2$, whereas the other one holds for $m > 1$.

Lemma 1.1.4 ([DiB93, Lemma I.4.4]). *Let $N \in \mathbb{N}$ and $m \in [2, \infty)$. There exists $\gamma > 0$ such that*

$$(a - b) \cdot (|a|^{m-2} a - |b|^{m-2} b) \geq \gamma |a - b|^m \tag{1.1.2}$$

for all a and b in $\mathbb{R}^N$.

Proof. For $m = 2$, the estimate is trivial for $\gamma = 1$. Hence we may assume $m > 2$ for the proof.

Then, by the chain rule, the function $f(x) := |x|^{m-2} = (|x|^2)^{m/2-1}$ is differentiable on $\mathbb{R}^N$ with derivative $\nabla f(x) = (m-2)|x|^{m-4} x^T$. Hence, by the fundamental theorem of calculus, the product rule, and the chain rule,

$$
\begin{aligned}
(a-b)(|a|^{m-2}a-|b|^{m-2}b) &= \int_0^1 \frac{\mathrm{d}}{\mathrm{d}s}\Big(\big|sa+(1-s)b\big|^{m-2}\big(sa+(1-s)b\big)\Big)\cdot(a-b)\,\mathrm{d}s \\
&= \int_0^1 \big|sa+(1-s)b\big|^{m-2}\big|a-b\big|^2\,\mathrm{d}s \\
&\quad + (m-2)\int_0^1 \big|sa+(1-s)b\big|^{m-4}(sa+(1-s)b \mid a-b)^2\,\mathrm{d}s
\end{aligned}
\tag{1.1.3}
$$

for all $a, b \in \mathbb{R}^N$. Since the second summand is positive,

$$
(a-b)\cdot(|a|^{m-2}a-|b|^{m-2}b) \geq \int_0^1 \big|sa+(1-s)b\big|^{m-2}\big|a-b\big|^2\mathrm{d}s \tag{1.1.4}
$$

for all a and b in $\mathbb{R}^N$.

Let a and b be arbitrary vectors in $\mathbb{R}^N$. We distinguish two cases. If $|a| \geq |b-a|$, then

$$
|sa+(1-s)b| \geq |a|-(1-s)|a-b| \geq s|a-b|,
$$

hence (1.1.2) follows from (1.1.4) with $\gamma = \int_0^1 s^{m-2}\mathrm{d}s = \frac{1}{m-1}$.

So now assume that $|a| < |b-a|$. Then

$$
|sa+(1-s)b| \leq (1-s)|a-b|+|a| \leq (2-s)|a-b|,
$$

hence

$$
\begin{aligned}
|a-b|^2\int_0^1 |sa+(1-s)b|^{m-2}\mathrm{d}s &\geq \int_0^1 \frac{|sa+(1-s)b|^m}{(2-s)^2}\mathrm{d}s \\
&\geq \frac{1}{4}\Big(\int_0^1 |sa+(1-s)b|^2\mathrm{d}s\Big)^{m/2}
\end{aligned}
$$

by Hölder's inequality. Thus, by (1.1.4),

$$
\begin{aligned}
(a-b)\cdot(|a|^{m-2}a-|b|^{m-2}b) &\geq \frac{1}{4}\Big(\int_0^1 |sa+(1-s)b|^2\mathrm{d}s\Big)^{m/2} \\
&= \frac{1}{4\cdot 3^{m/2}}\big(|a|^2+a\cdot b+|b|^2\big)^{m/2}.
\end{aligned}
\tag{1.1.5}
$$

Finally, since by Cauchy's inequality and Young's inequality we have $2|a\cdot b| \leq |a|^2+|b|^2$ and hence

$$
|a|^2+a\cdot b+|b|^2 \geq \frac{1}{2}(|a|^2+|b|^2) \geq \frac{1}{4}(|a|^2-2a\cdot b+|b|^2) = \frac{1}{4}|a-b|^2,
$$

we obtain from (1.1.5) that (1.1.2) holds for $|a| < |b-a|$ with $\gamma = \frac{1}{4\cdot 3^{m/2}\cdot 4^{m/2}}$.

Thus taking γ to be the smaller of the constants in the two cases, we deduce that (1.1.2) holds for all $a, b \in \mathbb{R}^N$. □

Lemma 1.1.5 ([Lin90, Appendix]). *Let $N \in \mathbb{N}$ and $m \in (1, 2]$. There exists $\gamma > 0$ such that*

$$(a-b)\cdot\left(|a|^{m-2}a - |b|^{m-2}b\right) \geq \gamma\left(|a|+|b|\right)^{m-2}|a-b|^2 \qquad (1.1.6)$$

for all a and b in $\mathbb{R}^N$.

Proof. For all a and b in $\mathbb{R}^N$ we have $|a-b|^2 \leq (|a|+|b|)^2$, hence

$$(|a|+|b|)^{m-2}|a-b|^2 \leq (|a|+|b|)^m.$$

Thus the right hand side of (1.1.6) can be continuously extended into $a = b = 0$ by zero. By continuity, it suffices to show the estimate for a and b in a dense subset of $\mathbb{R}^N \times \mathbb{R}^N$. Thus for the proof we can assume that zero is not on the line segment joining a and b.

In this situation, (1.1.3) remains valid. Thus, by the Cauchy-Schwarz inequality,

$$(a-b)\cdot\left(|a|^{m-2}a - |b|^{m-2}b\right) \geq (m-1)\int_0^1 |sa+(1-s)b|^{m-2}\mathrm{d}s\,|a-b|^2.$$

Since

$$|sa+(1-s)b|^{m-2} \geq (|a|+|b|)^{m-2}$$

for all $s \in [0,1]$, estimate (1.1.6) is satisfied for $\gamma = m-1$. □

1.2 Function spaces

In this section, we lay down the notation for various spaces of functions and their corresponding norms. Moreover, we recall some basic rules of real calculus for such functions. More information on Sobolev and Hölder spaces can be found in any textbook about partial differential equations. For a more detailed analysis of Sobolev functions, we refer to Ziemer [Zie89]. More information about geometric measure theory, like the change of variable formula, can be found in the books by Federer [Fed69] or Evans and Gariepy [EG92].

Definition 1.2.1 (Continuously differentiable functions). Let $\Omega \subset \mathbb{R}^N$ be an open set. For $k \in \mathbb{N}_0$, we denote by $\mathrm{C}^k(\overline{\Omega})$ the set of functions u in $\mathrm{C}^k(\Omega)$ such that u and all of its partial derivatives of order up to k have continuous extensions to $\overline{\Omega}$. For $u \in \mathrm{C}^k(\overline{\Omega})$ we say that u is *k times continuously differentiable on the topological closure $\overline{\Omega}$ of Ω* or that u is *k times continuously differentiable up to the topological boundary $\partial\Omega$ of Ω*. For such functions we do not distinguish between u itself and its continuous extension to $\overline{\Omega}$. Moreover, we set $\mathrm{C}^\infty(\overline{\Omega}) := \bigcap_{k=0}^\infty \mathrm{C}^k(\overline{\Omega})$. We say that a (continuous) function u has *compact support in Ω* if the set $\{u \neq 0\}$ is relatively compact in the relative topology of Ω, which is the same as to say that u vanishes outside a compact subset of Ω. We write

$C_c^k(\Omega)$, $k \in \mathbb{N}_0 \cup \{\infty\}$, for the functions in $C^k(\Omega)$ that have compact support in Ω. In particular, $C_c^k(\Omega)$ is a subset of $C^k(\overline{\Omega})$.

It is clear from the definition that if ω is an open subset of Ω, then for every $k \in \mathbb{N}_0 \cup \{\infty\}$ the restriction $u|_\omega$ of functions u in $C^k(\overline{\Omega})$ to ω is in $C^k(\overline{\omega})$.

Definition 1.2.2 (Hölder continuous functions). Let $\Omega \subset \mathbb{R}^N$ be open and bounded, and let $\alpha \in (0,1]$. We say that a function u on Ω is *Hölder continuous with exponent α and Hölder constant $c \geq 0$* if $|u(x) - u(y)| \leq c|x-y|^\alpha$ for all $x, y \in \Omega$. Then $u \in C(\overline{\Omega})$, and in particular u is bounded. The set $C^{0,\alpha}(\Omega)$ of all Hölder continuous functions on Ω with exponent α, equipped with the norm

$$\left\|u\right\|_{C^{0,\alpha}(\Omega)} := \left\|u\right\|_{L^\infty(\Omega)} + \sup_{x \neq y} \frac{|u(x) - u(y)|}{|x-y|^\alpha},$$

is a Banach space. Functions in $C^{0,1}(\Omega)$ are also called *Lipschitz continuous.*

Theorem 1.2.3 (Change of variables, [Fed69, Theorems 3.1.6 and 3.2.5]). *Let E and F be a measurable subsets of $\mathbb{R}^N$, let T be a Lipschitz continuous bijective mapping from E to F, and let u be a integrable function on F. Then T is differentiable almost everywhere, and*

$$\int_F u = \int_E u(Tx)\,|\det(T'(x))|\,dx,$$

where in particular the integral on the right hand side exists.

Definition 1.2.4 (Weakly differentiable functions). Let $\Omega \subset \mathbb{R}^N$ be open. We say that a function $u \in L^1_{\text{loc}}(\Omega)$ is *weakly differentiable* if there exist functions v_j, $j = 1, \dots, N$, in $L^1_{\text{loc}}(\Omega)$ such that v_j is the distributional derivative of u with respect to x_j, i.e.,

$$\int_\Omega u \frac{\partial \varphi}{\partial x_j} = -\int_\Omega v_j \varphi \text{ for all } \varphi \in C_c^\infty(\Omega).$$

In this case we write $D_j u := v_j$ and call the row vector $\nabla u := (D_1 u, \dots, D_N u)$ the *weak derivative* or *(weak) gradient* of u. The set of all functions u in $L^1_{\text{loc}}(\Omega)$ that are weakly differentiable is denoted by $W^{1,1}_{\text{loc}}(\Omega)$. Every function in $u \in C^1(\overline{\Omega})$ is weakly differentiable, and we often write $D_j u$ also for the classical derivatives, which coincide with the weak derivatives. We write $W^{1,p}(\Omega)$ for the set of all weakly differentiable functions u in $L^p(\Omega)$ such that ∇u is in $L^p(\Omega)$. Equipped with the norm

$$\left\|u\right\|_{W^{1,p}(\Omega)} := \begin{cases} \left(\|u\|^p_{L^p(\Omega)} + \sum_{j=1}^N \|D_j u\|^p_{L^p(\Omega)}\right)^{1/p}, & p \in [1,\infty), \\ \max\left\{\|u\|_{L^\infty(\Omega)}, \|D_1 u\|_{L^\infty(\Omega)}, \dots, \|D_N u\|_{L^\infty(\Omega)}\right\}, & p = \infty, \end{cases}$$

the *Sobolev space* $W^{1,p}(\Omega)$ is a Banach space, and for $p = 2$ even a Hilbert space for the obvious scalar product. If $1 < p < \infty$, the space $W^{1,p}(\Omega)$ is reflexive. The closure of $C_c^\infty(\Omega)$ in $W^{1,p}(\Omega)$ is denoted by $W_0^{1,p}(\Omega)$. If u is a function on Ω such that every point in Ω admits an open neighborhood $\Omega' \subset \Omega$ such that $u \in W^{1,p}(\Omega')$, we say that u is *locally in* $W^{1,p}$ and write $u \in W^{1,p}_{\mathrm{loc}}(\Omega)$.

Theorem 1.2.5 (Chain rule, [Zie89, Theorem 2.2.2]). *Let U and V be open subsets of $\mathbb{R}^N$, let u be a function in $W^{1,p}(V)$, $p \in [1,\infty]$, and let $T\colon U \to V$ be a bi-Lipschitz function, i.e., T is bijective from U to V and T and its inverse T^{-1} are Lipschitz continuous functions. Then T is differentiable almost everywhere, $u \circ T$ is in $W^{1,p}(U)$, and*

$$\nabla(u \circ T) = \big(\nabla u \circ T\big)T'$$

almost everywhere.

Lemma 1.2.6 ([Zie89, Theorem 2.1.11]). *Let $\Omega \subset \mathbb{R}^N$ be a bounded open set. Let $f\colon \mathbb{R} \to \mathbb{R}$ be Lipschitz continuous and $u \in W^{1,p}(\Omega)$, $p \in [1,\infty]$. Then $f \circ u \in W^{1,p}(\Omega)$ with weak derivative*

$$\nabla(f \circ u) = (f' \circ u)\nabla u \tag{1.2.1}$$

almost everywhere.

Lemma 1.2.7. *Let (Ω, μ) be a σ-finite measure space. Let $u\colon [0,1] \times \Omega \to \mathbb{R}^N$ be measurable. Assume that $u(\cdot, x)$ is continuously differentiable on $[0,1]$ for almost every $x \in \Omega$, and assume that there exists $\psi \in L^1(\Omega)$ such that the partial derivative u_t of u with respect to t satisfies $|u_t(t,x)| \le \psi(x)$ for all $t \in [0,1]$ and almost every $x \in \Omega$. Let g be a measurable function on $\Omega \times \mathbb{R}^N$ such that there exists $L \ge 0$ with the property that for almost every $x \in \Omega$ the function $z \mapsto g(x,z)$ is Lipschitz continuous on $\mathbb{R}^N$ with Lipschitz constant at most L. Then the integral*

$$I(t) := \int_\Omega g(x, u(t,x))\,\mathrm{d}x$$

exists for all $t \in [0,1]$ and I is a function in $W^{1,1}(0,1)$ with weak derivative

$$I'(t) = \int_\Omega \nabla_z g(x, u(t,x))\; u_t(t,x)\,\mathrm{d}x$$

for almost every $t \in [0,1]$.

Proof. For almost every $x \in \Omega$, the function $g(x, u(\cdot, x))$ is Lipschitz continuous, and its derivative equals

$$w(t,x) := \nabla_z g(x, u(t,x))\; u_t(t,x)$$

for almost every $t \in [0,1]$. Thus

$$g(x, u(t,x)) = g(x, u(0,x)) + \int_0^t w(s,x)\,\mathrm{d}s \tag{1.2.2}$$

for every $t \in [0,1]$ and almost every $x \in \Omega$. By assumption, $|w(x,t)| \le L\psi(x)$ for all $t \in [0,1]$ and almost every $x \in \Omega$. Hence w is integrable over $[0,1] \times \Omega$ with respect to the product measure. Integrating (1.2.2) over Ω and applying Fubini's theorem, we obtain that

$$I(t) = \int_\Omega g(x, u(0,x))\mathrm{d}x + \int_0^t \int_\Omega w(s,x)\mathrm{d}x\,\mathrm{d}s$$

for all $t \in [0,1]$. Thus the integral in the definition of $I(t)$ exists and I is in $W^{1,1}(0,1)$ with weak derivative

$$I'(t) = \int_\Omega w(t,x)\mathrm{d}x$$

for almost every $t \in [0,1]$. □

1.3 Lipschitz domains

It is well understood that many properties of solutions of partial differential equations, in particular their regularity, depend on the smoothness of the boundary of the underlying Euclidean domain. For applications, the most important class of domains are those with Lipschitz boundary, and in particular polygons. Fortunately, several important theorems for smooth domains remain valid for Lipschitz domains, which makes it possible to develop a satisfactory theory in this setting. Here we quote some results that will be needed later.

Definition 1.3.1. A subset Ω of $\mathbb{R}^N$ is called a *(bounded) Lipschitz domain* if Ω is open and bounded and for every $z \in \partial\Omega$ there exist an orthogonal matrix $\mathcal{O}$, a radius $r > 0$, and a Lipschitz continuous real-valued function ψ on $\mathbb{R}^{N-1}$ with $\psi(0) = 0$ such that for the neighborhood

$$G := \big\{(y, \psi(y) + s) : y \in B(0,r) \subset \mathbb{R}^{N-1}, s \in (-r, r)\big\}$$

of zero we are in the situation that

$$\mathcal{O}(\Omega - z) \cap G = \big\{(y, \psi(y) + s) : y \in B(0,r) \subset \mathbb{R}^{N-1}, s \in (0, r)\big\}.$$

Thus, colloquially speaking, an open, bounded subset Ω of $\mathbb{R}^N$ is a Lipschitz domain if locally, after rotation, it can be represented as the epigraph of a Lipschitz continuous function on $\mathbb{R}^{N-1}$.

Remark 1.3.2. Similarly, we say that $\Omega \subset \mathbb{R}^N$ is of class C^k, $k \in \mathbb{N}_0$, if in Definition 1.3.1 we require ψ to be in $\mathrm{C}^k(\mathbb{R}^{N-1})$ instead of being Lipschitz continuous.

Definition 1.3.3. Let Ω and Ω' be open subsets of $\mathbb{R}^N$. We say that Ω' is *relatively compact in* Ω if Ω' is relatively compact in Ω with respect to the relative topology of Ω, i.e., if there exists a compact set K such that $\Omega' \subset K \subset \Omega$. This is equivalent to saying that Ω' is a bounded subset of Ω whose closure is still contained in Ω.

Definition 1.3.4 (Outer normal). Let $\Omega \subset \mathbb{R}^N$ be a bounded Lipschitz domain. Fix $z \in \partial\Omega$ and let r, $\mathcal{O}$, G, and ψ be defined as in Definition 1.3.1. For $y \in B(0,r)$ we set

$$\nu\Big(\mathcal{O}^{-1}(y,\psi(y)) + z\Big) := \begin{cases} \frac{1}{\left(|\nabla\psi(y)|^2+1\right)^{1/2}}\ \mathcal{O}^{-1}\begin{pmatrix}\nabla\psi(y)^T\\ -1\end{pmatrix} & \text{if } \psi \text{ is differentiable at } y,\\ 0 & \text{otherwise.}\end{cases}$$

Since this definition is independent of the choice of local coordinates, we have in this way defined a function $\nu\colon \partial\Omega \to \mathbb{R}^N$ with $|\nu(z)| \le 1$ for all $z \in \partial\Omega$. This function ν is called the *outer (unit) normal of* Ω.

Geometrically, $\nu(z)$ is the normalized vector that is orthogonal to the tangent plane to Ω at z wherever a unique tangent plane exists, pointing outwards from Ω. This geometric interpretation makes sense only because Ω, being a Lipschitz domain, is not too ill-behaved at the boundary. Nevertheless, for general domains, there is still a sensible notion of an exterior normal [Fed69, §4.5.5], but we will not need this here.

Theorem 1.3.5 (Divergence theorem, [EG92, §4.3]). *Let Ω be a bounded Lipschitz domain. Then there exists a unique finite Borel measure σ on $\partial\Omega$ such that*

$$\int_\Omega u \,\operatorname{div} h = \int_{\partial\Omega} u\,(h\cdot\nu)\,\mathrm{d}\sigma - \int_\Omega \nabla u\, h \tag{1.3.1}$$

holds for all $u \in \mathrm{C}^1(\overline{\Omega})$ and $h \in \mathrm{C}^1(\overline{\Omega};\mathbb{R}^N)$, where $\operatorname{div} h := \sum_{j=1}^N D_j h_j$ denotes the divergence *of the vector field h and ν is the outer normal of Ω. We call σ the* surface measure of Ω. *We agree that integrals over $\partial\Omega$ are always to be understood to be taken with respect to the surface measure and usually suppress it in the notation of boundary integrals. As a consequence of* (1.3.1), *we obtain the rule*

$$\int_\Omega u\, D_j v = \int_{\partial\Omega} uv\,\nu_j - \int_\Omega D_j u\, v \tag{1.3.2}$$

for u and v in $\mathrm{C}^1(\overline{\Omega})$.

Theorem 1.3.6 (Approximation by smooth functions, [EG92, §4.2.1], [Maz85, §1.1.6]). *Let Ω be a bounded domain with continuous boundary, for example a bounded Lipschitz domain. Then $\mathrm{C}^\infty(\overline{\Omega})$ is dense in $W^{1,p}(\Omega)$ for every $p \in [1,\infty)$.*

Theorem 1.3.7 (Trace operator, [EG92, §4.3]). *Let Ω be a bounded Lipschitz domain. Then for every $p \in [1,\infty]$, there exists a continuous linear operator T from $W^{1,p}(\Omega)$ to $L^p(\partial\Omega)$ called the* trace operator *such that $Tu = u|_{\partial\Omega}$ for all u in $\mathrm{C}(\overline{\Omega}) \cap W^{1,p}(\Omega)$. We usually write $u|_{\partial\Omega} := Tu$ for $u \in W^{1,p}(\Omega)$.*

Since $\mathrm{C}(\overline{\Omega}) \cap W^{1,p}(\Omega)$ is dense in $W^{1,p}(\Omega)$ by Theorem 1.3.6, the trace operator in Theorem 1.3.7 is unique.

Remark 1.3.8. By density of $C^1(\overline{\Omega})$ and continuity of the trace operator, an approximation argument shows that (1.3.1) and (1.3.2) remain true if $u \in W^{1,p}(\Omega)$, $h \in W^{1,p'}(\Omega; \mathbb{R}^N)$, and $v \in W^{1,p'}(\Omega)$ for some $p \in [1, \infty]$.

The following two special cases of the Sobolev embedding theorems are widely known. They are contained among the results in Section 1.4 of Grisvard's book [Gri85].

Theorem 1.3.9. *Let Ω be a bounded Lipschitz domain, and let $p \in [1, \infty)$. Then $W^{1,p}(\Omega)$ embeds continuously into $L^{Np/(N-p)}(\Omega)$ if $p < N$, into $L^r(\Omega)$ for every $r < \infty$ if $p = N$, and into $C^{0,\alpha}(\Omega)$ for some $\alpha \in (0,1)$ if $p > N$. The latter two embeddings are compact (possibly only after shrinking α). For $p < N$, the space $W^{1,p}(\Omega)$ embeds compactly into $L^r(\Omega)$ for every $r < \frac{Np}{N-p}$.*

Theorem 1.3.10. *Let Ω be a bounded Lipschitz domain, and let $p \in [1, \infty)$. The trace operator is continuous from $W^{1,p}(\Omega)$ to $L^{(N-1)p/(N-p)}(\partial\Omega)$ if $p < N$, to $L^r(\partial\Omega)$ for every $r < \infty$ if $p = N$, and to $C(\partial\Omega)$ if $p > N$. The operator is compact in the latter two cases. For $p < N$, it is compact from $W^{1,p}(\Omega)$ to $L^r(\partial\Omega)$ for every $r < \frac{(N-1)p}{N-p}$.*

In the limit case as $p = \infty$ and $\alpha = 1$, the embedding in Theorem 1.3.9 becomes even an equality of spaces.

Proposition 1.3.11. *If Ω is a bounded Lipschitz domain, then $W^{1,\infty}(\Omega) = C^{0,1}(\Omega)$, and the weak derivative equals the classical derivative almost everywhere.*

Proof. We have that $W^{1,\infty}(\mathbb{R}^N) = C^{0,1}(\mathbb{R})$ [EG92, §6.2], that every function in $W^{1,\infty}(\Omega)$ has an extension to a function in $W^{1,\infty}(\mathbb{R}^N)$ [Ste70, §3.1], and that for every Lipschitz continuous function on Ω there exists a Lipschitz continuous extension to $\mathbb{R}^N$ [Fed69, Theorem 2.10.43]. Since the restriction of a Lipschitz continuous function is Lipschitz continuous and the restriction of a function in $W^{1,\infty}(\mathbb{R}^N)$ to Ω is in $W^{1,\infty}(\Omega)$, this proves the claim. □

The following lemma constitutes a nice tool for later application. Daners [Dan09] used this very elementary observation to show that for partial differential equations with Robin boundary conditions on Lipschitz domains it is no loss of generality to assume that the Robin term is positive. We will use this argument in Lemma 4.1.4. However, the main application of this lemma in the proof of Lemma 4.1.6 will be completely unrelated to this argument. Since the result is not common knowledge, we include the proof.

Lemma 1.3.12 ([Dan09, Lemma 3.2]). *Let Ω be a bounded Lipschitz domain with outer normal ν. There exists a smooth vector field $h \in C^\infty(\mathbb{R}^N; \mathbb{R}^N)$ such that $h \cdot \nu \geq 1$ almost everywhere on $\partial\Omega$.*

Proof. By compactness of $\partial\Omega$ there exist finitely many z_j such that the open sets $U_j := \mathcal{O}_j^{-1} G_j + z_j$ defined as in Definition 1.3.1 cover $\partial\Omega$. For fixed j, we define h_j on $U_j \cap \partial\Omega$ by

$$h_j\left(\mathcal{O}_j^{-1}(y, \psi_j(y)) + z_j\right) := \mathcal{O}_j^{-1} \begin{pmatrix} 0 \\ -1 \end{pmatrix} \quad \text{for } y \in B(0, r_j).$$

Then, since $\mathcal{O}_j$ is orthogonal and since all the functions $\nabla\psi_j$ are bounded, there exists $\varepsilon_0 > 0$ such that

$$(h_j \cdot \nu)\big(\mathcal{O}^{-1}(y, \psi(y)) + z\big) = \frac{1}{\big(|\nabla\psi_j(y)|^2 + 1\big)^{1/2}} \geq \varepsilon_0,$$

i.e., $h_j \cdot \nu \geq \varepsilon_0$ almost everywhere on $U_j \cap \partial\Omega$.

Let ψ_j be a partition of unity subordinate to U_j, $\psi_j \in C_c^\infty(\mathbb{R}^N; \mathbb{R}^N)$, compare for example [Zie89, Lemma 2.3.1]. Then $h := \sum_j \psi_j h_j$ is a smooth vector field such that

$$h \cdot \nu = \sum\nolimits_j \psi_j (h_j \cdot \nu) \geq \sum\nolimits_j \psi_j \varepsilon_0 = \varepsilon_0$$

almost everywhere on $\partial\Omega$. Thus $\varepsilon_0^{-1} h$ is a smooth vector field with the desired property. □

1.4 Functional analysis

In many respects, functional analysis provides the appropriate language for differential equations. The reader is assumed to be familiar with its basic concepts, and for example the closed graph theorem will be used freely. The collection of results in this section shall provide an overview over the author's terminology and introduce some results that might not be familiar to the reader.

Definition 1.4.1. Let X and Y be a Banach spaces.

(i) We say that A is a *linear operator* from X to Y if A is a linear mapping from a subspace $D(A)$ of X, called its *domain*, to the space Y.

(ii) We say that A is *densely defined* if $D(A)$ is dense in X.

(iii) We say that A is *bounded* if it maps subsets in $D(A)$ that are bounded with respect to the norm of X to bounded subsets of Y, or, equivalently, if A is continuous. Then there exists a unique bounded extension of A to the closure of $D(A)$, and we identify A with this extension.

(iv) We say that A is *closed* if the set $\{(x, Ax) : x \in D(A)\}$ is closed in the product space $X \oplus Y$.

(v) We say that A is *invertible* if A is bijective from $D(A)$ to Y and the inverse mapping A^{-1} is bounded.

(vi) Writing $\lambda + A$ for some $\lambda \in \mathbb{R}$, we refer to the linear operator with domain $D(A)$ that maps x to $\lambda x + Ax$.

(vii) We say that $\lambda \in \mathbb{R}$ is in the *resolvent set* $\rho(A)$ of A if $\lambda - A$ is invertible, and we frequently write $R(\lambda, A)$ instead of $(\lambda - A)^{-1}$.

In the definition of invertibility, we do not require A itself to be bounded. Every invertible operator is closed, and by the closed graph theorem, a closed operator is invertible if and only it is bijective.

Lemma 1.4.2. *Let (Ω, μ) be a measure space and $p, q \in [1, \infty]$. Then the intersection of the closed unit ball of $L^p(\Omega)$ with $L^q(\Omega)$ is a closed subset of $L^q(\Omega)$.*

Proof. Let (u_n) be a sequence in $L^q(\Omega)$ that lies in the closed unit ball of $L^p(\Omega)$ and converges to $u \in L^q(\Omega)$ in the norm of $L^q(\Omega)$. Passing to a subsequence, we can assume that (u_n) converges to u pointwise almost everywhere on Ω. If $p = \infty$, this shows that $|u| \leq 1$ almost everywhere, i.e., that u is in the closed unit ball of $L^p(\Omega)$. If $p \in [1, \infty)$, by Fatou's lemma

$$\int_\Omega |u|^p = \int_\Omega \left(\liminf_{n\to\infty} |u_n|\right)^p \leq \liminf_{n\to\infty} \int_\Omega |u_n|^p \leq 1,$$

showing that u is in the closed unit ball of $L^p(\Omega)$. □

Lemma 1.4.3. *Let (Ω, μ) be a measure space, let p and q be in $[1, \infty)$, and let $r \geq 0$. Assume $\frac{1}{q} - \frac{r}{p} \in (0, 1)$ and let $h \colon \Omega \times \mathbb{R} \to \mathbb{R}$ be a measurable function that is continuous in its second argument and satisfies*

$$|h(\omega, u)| \leq \varphi(\omega)|u|^r + \psi(\omega)$$

for all $u \in \mathbb{R}$ and $\omega \in \Omega$ with non-negative functions $\varphi \in L^{pq/(p-qr)}(\Omega)$ and $\psi \in L^q(\Omega)$. Then the Nemitsky operator *$u \mapsto h(\cdot, u(\cdot))$ is well-defined and continuous from $L^p(\Omega)$ to $L^q(\Omega)$.*

Proof. Define $H(u) := h(\cdot, u(\cdot))$ for $u \in L^p(\Omega)$. Then $H(u)$ is measurable being a composition of measurable functions. Moreover,

$$\|H(u)\|_{L^q(\Omega)} \leq \|\varphi\|_{L^{pq/(p-qr)}(\Omega)} \||u|^r\|_{L^{p/r}(\Omega)} + \|\psi\|_{L^q(\Omega)} = c_1 \|u\|^r_{L^p(\Omega)} + c_2 \qquad (1.4.1)$$

for some positive constants c_1 and c_2 by Minkowski's and Hölder's inequalities. Thus $H(u) \in L^q(\Omega)$ for $u \in L^p(\Omega)$, proving that the Nemitsky operator is well-defined.

Now let (u_n) be a sequence in $L^p(\Omega)$ that converges in the norm of $L^p(\Omega)$ to $u \in L^p(\Omega)$. Passing to a subsequence, we can assume that (u_n) converges pointwise almost everywhere to u and that there exists $v \in L^p(\Omega)$ such that $|u_n| \leq v$ almost everywhere for all $n \in \mathbb{N}$. Then, by continuity of h, the functions $H(u_n)$ converge pointwise almost everywhere to $H(u)$, and $|H(u_n)| \leq \varphi v^r + \psi$, with the right hand side being in $L^q(\Omega)$ by (1.4.1). By Lebesgue's dominated convergence theorem this proves that $H(u_n)$ converges to $H(u)$ in $L^q(\Omega)$. In this way we see that every subsequence of $H(u_n)$ has a subsequence which converges to $H(u)$ in $L^q(\Omega)$ in $L^q(\Omega)$. Thus $H(u_n)$ converges to $H(u)$, proving that H is continuous. □

Theorem 1.4.4 ([Zie89, Theorem 4.3.3]). *Let Ω be an open subset of $\mathbb{R}^N$. Then for every bounded linear functional φ on $W_0^{1,p}(\Omega)$, $1 \le p < \infty$, there exist $h_0, h_1, \dots, h_N \in L^{p'}(\Omega)$ such that*

$$\varphi(\eta) = \int_\Omega \eta\, h_0 + \sum_{j=1}^N \int_\Omega D_j\eta\, h_j$$

holds for all $\eta \in W_0^{1,p}(\Omega)$.

Proof. Via the isometry $u \mapsto (u, D_1u, \dots, D_Nu)$ from $W_0^{1,p}(\Omega)$ to $L^p(\Omega)^{N+1}$ we can consider $W_0^{1,p}(\Omega)$ to be a closed subspace of $L^p(\Omega)^{N+1}$. With this identification, by the Hahn-Banach theorem there exists a bounded linear functional ψ on $L^p(\Omega)^{N+1}$ extending φ. Since the dual space of $L^p(\Omega)$ is $L^{p'}(\Omega)$, it is easy to deduce that

$$\psi\big(v_0, v_1, \dots, v_N\big) = \sum_{j=0}^N \int_\Omega v_j h_j$$

for some $h_j \in L^{p'}(\Omega)$. Restricting to $W_0^{1,p}(\Omega)$ and using the identification of functions in $W_0^{1,p}(\Omega)$ with elements of $L^p(\Omega)^{N+1}$, this is the claim. □

Definition 1.4.5. Let V be a real Hilbert space. A mapping a from $V \times V$ to $\mathbb{R}$ is called a *(bilinear) form* if a is *bilinear*, i.e., linear in both of its arguments. A form a is called *bounded* if there exists $M \ge 0$ such that $|a(u,v)| \le M\|u\|_V\|v\|_V$ for all u and v in V, i.e., if a is continuous. It is called *coercive* is there exists $\alpha > 0$ such that $a(u,u) \ge \alpha\|u\|_V^2$ for all $u \in V$. If V is continuously embedded into another Hilbert space H, then a is called *H-elliptic* if there exists $\omega \in \mathbb{R}$ such that the form $(u,v) \mapsto a(u,v) + \omega\,(u \mid v)_H$ is coercive.

The following theorem is classical. In text books the estimate that is contained in the statement is often not made explicit, even though it is implicitly proved along the way.

Theorem 1.4.6 (Lax-Milgram's lemma, [GT01, Theorem 5.8]). *Let H be a Hilbert space, and let $a\colon H \times H \to \mathbb{R}$ be a continuous, coercive form. Then there exists a constant c with the following property. If φ is a bounded linear functional on H, then there exists a unique $u \in H$ such that $a(u,v) = \varphi(v)$ for all $v \in H$, and this u satisfies $\|u\|_H \le c\|\varphi\|_{H'}$.*

Lemma 1.4.7. *Let X be a reflexive Banach space, and let Y and Z be Banach spaces such that X is continuously embedded into Z. Moreover, let $T\colon X \to Y$ be a linear, compact operator. Then for every $\varepsilon > 0$ there exists $c_\varepsilon \ge 0$ such that*

$$\|Tx\|_Y \le \varepsilon\|x\|_X + c_\varepsilon\|x\|_Z$$

for all $x \in X$.

Proof. Assume for contradiction that there exists $\varepsilon_0 > 0$ such that for every n we can find $x_n \in X$ satisfying

$$\|Tx_n\|_Y > \varepsilon_0\|x_n\|_X + n\|x_n\|_Z. \tag{1.4.2}$$

Rescaling x_n we can without loss of generality assume that $\|x_n\|_X = 1$ for all $n \in \mathbb{N}$. By reflexivity, after passing to a subsequence we can assume that $x_n \rightharpoonup x$ in X for some $x \in X$, and thus in particular $x_n \rightharpoonup x$ in Z. Then by compactness of T we obtain that $Tx_n \to Tx$ in Y, and in particular that (Tx_n) is bounded in Y. Thus by (1.4.2)

$$\|x_n\|_Z < \frac{1}{n}\|Tx_n\|_Y \to 0 \quad (n \to \infty).$$

Since $x_n \rightharpoonup x$ in Z, this implies that $x = 0$. Hence $Tx_n \to 0$ in Y, contradicting the fact that by (1.4.2) we have $\liminf_{n\to\infty} \|Tx_n\|_Y \geq \varepsilon_0 > 0$. □

We will only need the following two special cases of Lemma 1.4.7, see also Theorems 1.3.9 and 1.3.10.

Example 1.4.8. Let $\Omega \subset \mathbb{R}^N$ be a bounded Lipschitz domain, $m \in (1, \infty)$, and $r \in (1, \infty)$. If $m < N$, assume in addition that $1 < r < \frac{Nm}{N-m}$. Then for every $\varepsilon > 0$ there exists $c_\varepsilon \geq 0$ such that

$$\|u\|_{L^r(\Omega)} \leq \varepsilon\|\nabla u\|_{L^m(\Omega)} + c_\varepsilon\|u\|_{L^m(\Omega)}$$

for every $u \in W^{1,m}(\Omega)$.

Example 1.4.9. Let $\Omega \subset \mathbb{R}^N$ be a bounded Lipschitz domain, $m \in (1, \infty)$, and $r \in (1, \infty)$. If $m < N$, assume in addition that $1 < r < \frac{(N-1)m}{N-m}$. Then for every $\varepsilon > 0$ there exists $c_\varepsilon > 0$ such that

$$\|u|_{\partial\Omega}\|_{L^r(\partial\Omega)} \leq \varepsilon \left\|\nabla u\right\|_{L^m(\Omega)} + c_\varepsilon \left\|u\right\|_{L^m(\Omega)}$$

for every $u \in W^{1,m}(\Omega)$.

1.5 Monotone operators

For solving non-linear equations, only few basic techniques are known, in particular when it comes to infinite dimensional spaces. A particularly powerful approach for partial differential equations is based on a monotonicity assumption on the operator, or, more generally, on the assumption that the operator is pseudo-monotone. We introduce the necessary notions in this section. Much more information can be found in Showalter's book [Sho97].

Definition 1.5.1. Let V be a Banach space and $\mathcal{A}$ be a (not necessarily linear) operator from V to the dual space V'. We call $\mathcal{A}$ *bounded* if it maps bounded subsets of V to bounded subsets of V'. We call $\mathcal{A}$ *monotone* if

$$\langle \mathcal{A}(u) - \mathcal{A}(v), u - v\rangle \geq 0$$

for all $u, v \in V$. We call $\mathcal{A}$ *strictly monotone* if

$$\langle \mathcal{A}(u) - \mathcal{A}(v), u - v \rangle > 0$$

whenever $u \neq v$. We call $\mathcal{A}$ *uniformly monotone* if there exist $\alpha > 0$ and $p > 1$ such that

$$\langle \mathcal{A}(u) - \mathcal{A}(v), u - v \rangle \geq \alpha \|u - v\|_V^p$$

for all $u, v \in V$. We say that $\mathcal{A}$ is *coercive* if

$$\lim_{\|v\|_V \to \infty} \frac{\langle \mathcal{A}(v), v \rangle_{V',V}}{\|v\|_V} = \infty. \tag{1.5.1}$$

Definition 1.5.2 (Pseudo-monotone operator)**.** Let V be a Banach space. An operator $\mathcal{A}\colon V \to V'$ is called *pseudo-monotone* if for every sequence (u_n) satisfying

$$u_n \rightharpoonup u \quad \text{and} \quad \limsup \langle \mathcal{A}(u_n), u_n - u \rangle_{V',V} \leq 0$$

we have that

$$\langle \mathcal{A}(u), u - v \rangle_{V',V} \leq \liminf \langle \mathcal{A}(u_n), u_n - v \rangle_{V',V}$$

for all $v \in V$.

Under a very mild additional regularity condition a monotone operator is pseudo-monotone. In this sense, pseudo-monotone operators are a generalization of monotone operators.

Proposition 1.5.3 ([Sho97, Proposition II.2.3])**.** *Let V be a Banach space and let $\mathcal{A}\colon V \to V'$ be a monotone operator. If in addition $\mathcal{A}$ is* hemicontinuous, *i.e., continuous along straight lines, then $\mathcal{A}$ is pseudo-monotone.*

Unlike the class of monotone operators, the class of pseudo-monotone operators is rather stable under small perturbations.

Theorem 1.5.4. *Let $\mathcal{A}\colon V \to V'$ be pseudo-monotone and let $\mathcal{B}\colon V \to V'$ be completely continuous, i.e., assume that $u_n \rightharpoonup u$ in V implies that $\mathcal{B}(u_n) \to \mathcal{B}(u)$ in V'. Then $\mathcal{A}+\mathcal{B}$ is pseudo-monotone.*

Proof. Let (u_n) be a sequence in V such that

$$u_n \rightharpoonup u \quad \text{and} \quad \limsup \langle \mathcal{A}(u_n) + \mathcal{B}(u_n), u_n - u \rangle_{V',V} \leq 0.$$

Since then $\mathcal{B}(u_n) \to \mathcal{B}(u)$, we have $\langle \mathcal{B}(u_n), u_n - u \rangle \to 0$ and hence

$$u_n \rightharpoonup u \quad \text{and} \quad \limsup \langle \mathcal{A}(u_n), u_n - u \rangle_{V',V} \leq 0.$$

By assumption, this implies

$$\langle \mathcal{A}(u), u - v\rangle_{V',V} \leq \liminf \langle \mathcal{A}(u_n), u_n - v\rangle_{V',V} \quad \text{for all } v \in V.$$

But since $\langle \mathcal{B}(u_n), u_n - v\rangle \to \langle \mathcal{B}(u), u - v\rangle$ for all $v \in V$ because of $\mathcal{B}(u_n) \to \mathcal{B}(u)$, this shows that

$$\langle \mathcal{A}(u) + \mathcal{B}(u), u - v\rangle_{V',V} \leq \liminf \langle \mathcal{A}(u_n) + \mathcal{B}(u_n), u_n - v\rangle_{V',V} \quad \text{for all } v \in V.$$

Thus $\mathcal{A} + \mathcal{B}$ is pseudo-monotone. □

Theorem 1.5.5 (Brézis, [Sho97, Theorem II.2.3]). *Let V be a separable, reflexive Banach space, let $\mathcal{A}\colon V \to V'$ be a bounded, pseudo-monotone, coercive operator. Then $\mathcal{A}$ is surjective.*

1.6 Interpolation theory

We start this section with a theorem that will be used to prove the continuity of certain nonlinear operators. The theorem and its corollary can be regarded as embedding results for an interpolation space between a Hölder space and a Lebesgue space, which is why it is included in this section. However, the proofs are direct and require no tools from interpolation theory.

Theorem 1.6.1 ([Lê07, Theorem 1.2]). *Let $\Omega \subset \mathbb{R}^N$ be a bounded Lipschitz domain, $\alpha \in (0, 1]$, and $q \in [1, \infty)$. Then there exist $c \geq 0$ and $\theta \in (0, 1)$ such that*

$$\|u\|_{L^\infty(\Omega)} \leq c \, \|u\|_{C^{0,\alpha}(\Omega)}^{\theta} \, \|u\|_{L^q(\Omega)}^{1-\theta} \tag{1.6.1}$$

for all $u \in C^{0,\alpha}(\Omega)$.

Proof. It is an easy consequence of Definition 1.3.1 that Ω satisfies an interior cone condition, i.e., that there exist a fixed opening angle ϑ and a positive number $r_0 > 0$ such that at every point x of $\overline{\Omega}$ we can find a truncated cone $K(x; r_0)$ with opening angle ϑ of height r_0 lying completely inside Ω.

Now let $u \in C^{0,\alpha}(\Omega)$, $u \neq 0$, be arbitrary and fix $x \in \overline{\Omega}$. Then

$$|u(x)| \leq \frac{|u(x) - u(y)|}{|x - y|^\alpha}|x - y|^\alpha + |u(y)| \leq \|u\|_{C^{0,\alpha}(\Omega)}|x - y|^\alpha + |u(y)|$$

for all $y \in \Omega$. Integrating over $K(x; r)$ for $r \in (0, r_0]$ with respect to y we obtain, with ω denoting the volume of $K(x; 1)$, that

$$\omega r^N |u(x)| \leq \|u\|_{C^{0,\alpha}(\Omega)} \int_0^r s^\alpha \cdot N\omega s^{N-1} \, \mathrm{d}s + (\omega r^N)^{1/q'} \|u\|_{L^q(\Omega)},$$

where we changed variables to radial coordinates and used Hölder's inequality. Thus

$$|u(x)| \leq c\left(r^{\alpha}\|u\|_{C^{0,\alpha}} + r^{-N/q}\|u\|_{L^q(\Omega)}\right) \tag{1.6.2}$$

for all $r \leq r_0$ with some constant $c \geq 0$ that is independent of u, x, and r. We pick $r_1 > 0$ such that $r_1^{\alpha+N/q} = \|u\|_{L^q(\Omega)}/\|u\|_{C^{0,\alpha}(\Omega)}$ and choose $\theta := \frac{N/q}{\alpha+N/q}$. If $r_1 \leq r_0$, we can insert $r = r_1$ into (1.6.2) and obtain

$$|u(x)| \leq 2cr_1^{-N/q}\|u\|_{L^q(\Omega)} = 2c\,\|u\|^{\theta}_{C^{0,\alpha}}\|u\|^{1-\theta}_{L^q(\Omega)}.$$

If, on the other hand, $r_1 > r_0$, i.e., $\|u\|_{L^q(\Omega)}/\|u\|_{C^{0,\alpha}(\Omega)} > r_0^{\alpha+N/q}$, then

$$|u(x)| \leq \|u\|_{C^{0,\alpha}(\Omega)} = \|u\|^{\theta}_{C^{0,\alpha}(\Omega)}\|u\|^{1-\theta}_{C^{0,\alpha}(\Omega)} \leq r_0^{-(1-\theta)(\alpha+N/q)}\|u\|^{\theta}_{C^{0,\alpha}(\Omega)}\|u\|^{1-\theta}_{L^q(\Omega)}.$$

Since in both cases the estimate of $u(x)$ is uniform with respect to x, we have shown that (1.6.1) holds. □

Corollary 1.6.2. *Let $\Omega \subset \mathbb{R}^N$ be a bounded Lipschitz domain, $0 < \beta < \alpha \leq 1$ and $q \in [1, \infty)$. Then there exist $\theta \in (0, 1)$ and $c \geq 0$ such that*

$$\|u\|_{C^{0,\beta}(\Omega)} \leq c\,\|u\|^{\theta}_{C^{0,\alpha}(\Omega)}\|u\|^{1-\theta}_{L^q(\Omega)} \tag{1.6.3}$$

for all $u \in C^{0,\alpha}(\Omega)$.

Proof. For u in $C^{0,\alpha}(\Omega)$ and $x \neq y$ in Ω we have

$$\frac{|u(x)-u(y)|}{|x-y|^{\beta}} = \left(\frac{|u(x)-u(y)|}{|x-y|^{\alpha}}\right)^{\beta/\alpha}|u(x)-u(y)|^{1-\beta/\alpha} \leq 2^{1-\beta/\alpha}\|u\|^{\beta/\alpha}_{C^{0,\alpha}(\Omega)}\|u\|^{1-\beta/\alpha}_{L^{\infty}(\Omega)}.$$

Hence, taking the supremum over all choices of x and y, for $c_0 := 2^{1-\beta/\alpha}$ and $\eta := \beta/\alpha$ we have

$$\|u\|_{C^{0,\beta}(\Omega)} \leq \|u\|_{L^{\infty}(\Omega)} + c_0\,\|u\|^{\eta}_{C^{0,\alpha}(\Omega)}\|u\|^{1-\eta}_{L^{\infty}(\Omega)} \leq (c_0+1)\|u\|^{\eta}_{C^{0,\alpha}(\Omega)}\|u\|^{1-\eta}_{L^{\infty}(\Omega)}$$

for all $u \in C^{0,\alpha}(\Omega)$. Now we can deduce from Theorem 1.6.1 that there exists a constant $c \geq 0$ such that

$$\|u\|_{C^{0,\beta}(\Omega)} \leq c\,\|u\|^{\eta+(1-\eta)\theta}_{C^{0,\alpha}(\Omega)}\|u\|^{(1-\eta)(1-\theta)}_{L^q(\Omega)}$$

for all $u \in C^{0,\alpha}(\Omega)$. This is (1.6.3). □

The goal of the remainder of this section is to obtain an interpolation result for nonlinear operators between Lebesgue spaces. In Chapter 3 this result will be needed for a bootstrapping argument.

Since the beautiful proof of this useful theorem seems to be available only in French and in order to be more self-contained, it is worthwhile to include a proof. We start by introducing real interpolation spaces via the K-method and proving some elementary properties that are needed. Then we prove the main theorem in this section, Theorem 1.6.9. Much more information about (linear) interpolation theory can be found in Triebel's book [Tri95].

Definition 1.6.3 (Real interpolation spaces, K-method)**.** Let A_0 and A_1 be Banach spaces.

(a) The couple $\{A_0, A_1\}$ is called an *interpolation couple* if A_0 and A_1 are both (continuously) embedded into a Hausdorff topological vector space $\mathscr{A}$. In this situation we can without loss of generality assume that $\mathscr{A} = A_0 + A_1$ with the usual norm, i.e.,

$$\|a\|_{A_0+A_1} := \inf\Big\{\|a_0\|_{A_0} + \|a_1\|_{A_1} : a = a_0 + a_1,\ a_0 \in A_0,\ a_1 \in A_1\Big\}$$

for $a \in A_0 + A_1$.

(b) If $\{A_0, A_1\}$ is an interpolation couple, we define the *K-functional* by

$$K(t,a) := K(t,a;A_0,A_1) := \inf\Big\{\|a_0\|_{A_0} + t\|a_1\|_{A_1} : a = a_0 + a_1,\ a_0 \in A_0,\ a_1 \in A_1\Big\}$$

for $a \in A_0 + A_1$ and $t \in (0,\infty)$.

(c) For $\theta \in (0,1)$ and $q \in [1,\infty]$, we define the real interpolation space $(A_0, A_1)_{\theta,q}$ by

$$(A_0,A_1)_{\theta,q} := \left\{a \in A_0 + A_1 : \|a\|_{(A_0,A_1)_{\theta,q}} := \left(\int_0^\infty t^{-\theta q} K(t,a)^q \frac{\mathrm{d}t}{t}\right)^{1/q} < \infty\right\}$$

if $q < \infty$ and by

$$(A_0,A_1)_{\theta,\infty} := \left\{a \in A_0 + A_1 : \|a\|_{(A_0,A_1)_{\theta,\infty}} := \sup_{0<t<\infty} t^{-\theta} K(t,a) < \infty\right\}$$

for $q = \infty$.

We collect and prove some properties of real interpolation spaces.

Lemma 1.6.4 ([Tri95, §1.3.3])**.** *Let $\{A_0, A_1\}$ be an interpolation couple, let $\theta \in (0,1)$, and let $q \in [1,\infty]$. Then $(A_0,A_1)_{\theta,q}$ is a Banach space which is continuously embedded into $A_0 + A_1$.*

Lemma 1.6.5 ([Tri95, §1.3.3])**.** *Let $A_0 \hookrightarrow A_1$ be Banach spaces, let $0 < \theta < \theta^* < 1$, and let $p, q \in [1,\infty]$. Then $(A_0,A_1)_{\theta,p}$ is continuously embedded into $(A_0,A_1)_{\theta^*,q}$.*

Lemma 1.6.6. *Let $\{A_0, A_1\}$ and $\{B_0, B_1\}$ be interpolation couples, let $\theta \in (0,1)$, and let $q \in [1,\infty]$. Then $\{A_0 \oplus B_0, A_1 \oplus B_1\}$ is an interpolation couple and*

$$(A_0,A_1)_{\theta,q} \oplus (B_0,B_1)_{\theta,q} = (A_0 \oplus B_0, A_1 \oplus B_1)_{\theta,q}$$

in the sense that the canonical identity from $(A_0+A_1)\oplus(B_0+B_1)$ onto $(A_0\oplus B_0)+(A_1\oplus B_1)$ induces an isomorphism between $(A_0,A_1)_{\theta,q} \oplus (B_0,B_1)_{\theta,q}$ and $(A_0 \oplus B_0, A_1 \oplus B_1)_{\theta,q}$.

Proof. If $\mathscr{A}$ and $\mathscr{B}$ are Hausdorff topological vector spaces such that A_0 and A_1 are continuously embedded into $\mathscr{A}$ and B_0 and B_1 are continuously embedded into $\mathscr{B}$, then the space $\mathscr{A} \times \mathscr{B}$ with the product topology is a Hausdorff topological vector space into which $A_0 \oplus B_0$ and $A_1 \oplus B_1$ are continuously embedded. Thus $\{A_0 \oplus B_0, A_1 \oplus B_1\}$ is an interpolation couple.

We equip $A_0 \oplus B_0$ and $A_1 \oplus B_1$ with the norms

$$\left\|(a,b)\right\|_{A_0 \oplus B_0} = \|a\|_{A_0} + \|b\|_{B_0} \text{ and } \left\|(a,b)\right\|_{A_1 \oplus B_1} = \|a\|_{A_1} + \|b\|_{B_1}.$$

Then we obtain from the definition of the K-functional that

$$K(t,(a,b); A_0 \oplus B_0, A_1 \oplus B_1) = K(t,a; A_0, B_0) + K(t,b; B_0, B_1)$$

for all $t \in (0,\infty)$, $a \in A_0 + A_1$, and $b \in B_0 + B_1$. Thus $(a,b) \in (A_0 \oplus B_0, A_1 \oplus B_1)_{\theta,q}$ if and only if $a \in (A_0, A_1)_{\theta,q}$ and $b \in (B_0, B_1)_{\theta,q}$, and in this case

$$\left\|(a,b)\right\|^q_{(A_0 \oplus B_0, A_1 \oplus B_1)_{\theta,q}} = \left\|a\right\|^q_{(A_0,A_1)_{\theta,q}} + \left\|b\right\|^q_{(B_0,B_1)_{\theta,q}}.$$

Hence the spaces are isomorphic. □

Theorem 1.6.7 (Real interpolation spaces, mean method [Tri95, §1.5]). *Let $\{A_0, A_1\}$ be an interpolation couple, $\theta \in (0,1)$, $1 \leq p, p_0, p_1 < \infty$, and*

$$\frac{1}{p} = \frac{1-\theta}{p_0} + \frac{\theta}{p_1}.$$

Then an element $a \in A_0 + A_1$ is in $(A_0, A_1)_{\theta,p}$ if and only if there exist measurable functions $a_0\colon (0,\infty) \to A_0$ and $a_1\colon (0,\infty) \to A_1$ such that

$$a = a_0(t) + a_1(t) \text{ for all } t \in (0,\infty) \tag{1.6.4}$$

and

$$M(a_0,a_1; A_0, A_1, p_0, p_1, \theta) := \left(\int_0^\infty t^{-\theta p_0} \|a_0(t)\|^{p_0}_{A_0} \frac{\mathrm{d}t}{t}\right)^{1/p_0} + \left(\int_0^\infty t^{(1-\theta)p_1} \|a_1(t)\|^{p_1}_{A_1} \frac{\mathrm{d}t}{t}\right)^{1/p_1}$$

is finite. In this case, for every $\varepsilon > 0$ there exists such a decomposition with the additional property that

$$\|a_0(t)\|_{A_0} + t\|a_1(t)\|_{A_1} \leq (1+\varepsilon)K(t,a)$$

holds for all $t \in (0,\infty)$.

Theorem 1.6.8 ([Tri95, §1.18.4]). *Let (Ω, μ) be a finite measure space. Fix exponents*

$1 \le p_0, p_1 \le \infty$ and an interpolation parameter $\theta \in (0,1)$, and let

$$\frac{1}{p} = \frac{1-\theta}{p_0} + \frac{\theta}{p_1}.$$

Then $(L^{p_0}(\Omega), L^{p_1}(\Omega))_{\theta,p} = L^p(\Omega)$.

We are now ready to prove the main result of this chapter, a nonlinear interpolation theorem, which is essentially due to Tartar [Tar72]. However, even tough all necessary ideas are included in his article, this specific version of his nonlinear interpolation results is not present there.

Theorem 1.6.9 (Nonlinear interpolation). *Let $A_0 \hookrightarrow A_1$ and $B_0 \hookrightarrow B_1$ be Banach spaces, where we regard A_0 and B_0 to be subspaces of A_1 and B_1, respectively. Let $T\colon A_1 \to B_1$ be a (not necessarily linear) operator that maps A_0 continuously into B_0. Assume that there exist exponents α in $(0,1]$ and $\beta \in (0,\infty)$ and functions c_0, c_1, and c_2 which are bounded on bounded sets such that*

$$\|Ta - Tb\|_{B_1} \le c_0(\|a\|_{A_1}, \|b\|_{A_1}) \|a-b\|_{A_1}^{\alpha} \text{ for all } a, b \in A_1 \tag{1.6.5}$$

and

$$\|Ta\|_{B_0} \le c_1(\|a\|_{A_1}) + c_2(\|a\|_{A_1}) \|a\|_{A_0}^{\beta} \text{ for all } a \in A_0. \tag{1.6.6}$$

Then for every $\theta \in (0,1)$ and $p \in [1,\infty)$, $p \ge \beta$, the operator T maps $(A_0, A_1)_{\theta,p}$ into $(B_0, B_1)_{\eta,q}$, where $\eta \in (0,1)$ and $q \in [1,\infty)$ are defined by

$$\frac{1-\eta}{\eta} = \frac{1-\theta}{\theta}\frac{\alpha}{\beta} \quad \text{and} \quad q = \frac{p}{(1-\eta)\beta + \eta\alpha}.$$

Proof. Let $a \in (A_0, A_1)_{\theta,p}$. By Theorem 1.6.7 we can find two measurable functions $a_0\colon (0,\infty) \to A_0$ and $a_1\colon (0,\infty) \to A_1$ with the properties that

$$a = a_0(t) + a_1(t) \text{ and } \|a_0(t)\|_{A_0} + t\|a_1(t)\|_{A_1} < 2K(t,a) \text{ for all } t \in (0,\infty) \tag{1.6.7}$$

and

$$M(a_0, a_1; p, p, \theta) = \left(\int_0^\infty t^{-\theta p} \|a_0(t)\|_{A_0}^p \frac{\mathrm{d}t}{t}\right)^{1/p} + \left(\int_0^\infty t^{(1-\theta)p} \|a_1(t)\|_{A_1}^p \frac{\mathrm{d}t}{t}\right)^{1/p} < \infty.$$

Since $A_0 \subset A_1$, $K(t,a) \le t\|a\|_{A_1}$ by the definition of the K-functional. Hence, by local boundedness of c_0, c_1, and c_2, there exist positive constants d_0, d_1, and d_2 that depend on a, but not on a_0 and a_1, such that

$$\|Ta - Ta_0(t)\|_{B_1} \le d_0 \|a_1(t)\|_{A_1}^{\alpha} \text{ for all } t \in (0,\infty) \tag{1.6.8}$$

and

$$\|Ta_0(t)\|_{B_0} \le d_1 + d_2 \|a_0(t)\|_{A_0}^{\beta} \text{ for all } t \in (0,\infty). \tag{1.6.9}$$

Define $\lambda := \frac{\eta}{\theta\beta} = \frac{1-\eta}{(1-\theta)\alpha} > 0$, $q_0 := \frac{p}{\beta} \in [1,\infty)$, $q_1 := \frac{p}{\alpha} \in [1,\infty)$, $b_0(t) := Ta_0(t^\lambda) \cdot \mathbb{1}_{\{t>1\}}$ and $b_1(t) := Ta - b_0(t)$. Since T is continuous from A_0 to B_0 and from A_1 to B_1, the functions b_0 and b_1 are measurable with values in B_0 and B_1, respectively. By (1.6.9),

$$\begin{aligned}
\Big(\int_0^\infty t^{-\eta q_0}\big\|b_0(t)\big\|_{B_0}^{q_0}\frac{\mathrm{d}t}{t}\Big)^{1/q_0} &= \Big(\int_1^\infty t^{-\eta q_0}\big\|Ta_0(t^\lambda)\big\|_{B_0}^{q_0}\frac{\mathrm{d}t}{t}\Big)^{1/q_0}\\
&= \Big(\int_1^\infty t^{-\eta q_0/\lambda}\big\|Ta_0(t)\big\|_{B_0}^{q_0}\frac{\lambda\,\mathrm{d}t}{t}\Big)^{1/q_0}\\
&\le d_1\Big(\int_1^\infty t^{-\theta p}\frac{\lambda\,\mathrm{d}t}{t}\Big)^{1/q_0} + d_2\Big(\int_0^\infty t^{-\theta p}\big\|a_0(t)\big\|_{A_0}^{\beta q_0}\frac{\lambda\,\mathrm{d}t}{t}\Big)^{1/q_0}\\
&\le d_1\Big(\frac{\lambda}{\theta p}\Big)^{1/q_0} + d_2\lambda^{1/q_0}M(a_0,a_1;p,p,\theta)^\beta < \infty
\end{aligned}$$

Similarly, by (1.6.8),

$$\begin{aligned}
&\Big(\int_0^\infty t^{(1-\eta)q_1}\big\|b_1(t)\big\|_{B_1}^{q_1}\frac{\mathrm{d}t}{t}\Big)^{1/q_1}\\
&\le \Big(\int_0^1 t^{(1-\eta)q_1}\big\|Ta\big\|_{B_1}^{q_1}\frac{\mathrm{d}t}{t}\Big)^{1/q_1} + \Big(\int_1^\infty t^{(1-\eta)q_1}\big\|Ta - Ta_0(t^\lambda)\big\|_{B_1}^{q_1}\frac{\mathrm{d}t}{t}\Big)^{1/q_1}\\
&= \Big(\frac{1}{(1-\eta)q_1}\Big)^{1/q_1}\big\|Ta\big\|_{B_1} + \Big(\int_1^\infty t^{(1-\eta)q_1/\lambda}\big\|Ta - Ta_0(t)\big\|_{B_1}^{q_1}\frac{\lambda\,\mathrm{d}t}{t}\Big)^{1/q_1}\\
&\le \Big(\frac{1}{(1-\theta)q_1}\Big)^{1/q_1}\|Ta\|_{B_1} + d_0\Big(\int_0^\infty t^{(1-\theta)p}\big\|a_1(t)\big\|_{A_1}^{\alpha q_1}\frac{\lambda\,\mathrm{d}t}{t}\Big)^{1/q_1}\\
&\le \Big(\frac{1}{(1-\theta)q_1}\Big)^{1/q_1}\|Ta\|_{B_1} + d_0\lambda^{1/q_1}M(a_0,a_1;p,p,\theta)^\alpha < \infty.
\end{aligned}$$

Thus, since $b_0(t) + b_1(t) = Ta$ for all t by definition and since

$$\frac{1}{q} = \frac{1-\eta}{q_0} + \frac{\eta}{q_1},$$

we obtain from Theorem 1.6.7 that $Ta \in (B_0, B_1)_{\eta,q}$. □

Remark 1.6.10. In the situation of Theorem 1.6.9, we obtain an estimate of the same kind as (1.6.6) also for the interpolation spaces, and we can even estimate the coefficients. In fact, taking the infimum of the functional M in Theorem 1.6.7 over all admissible decompositions, we arrive at an equivalent norm on $(A_0, A_1)_{\theta,p}$, and in the proof of Theorem 1.6.9 we estimated these expressions. Moreover, if the operator satisfies a Hölder condition like (1.6.5) at both ends of the interpolation scale, then it is locally Hölder continuous on the interpolation spaces as well. This can be proved by using Theorem 1.6.9 and a simple trick. For all these results, the reader is referred to Tartar's article [Tar72].

We can specialize this nonlinear interpolation result to linear operators, where the assumptions of Theorem 1.6.9 are fulfilled with $c_1 = 0$ and $\beta = \alpha = 1$ and where c_0 and c_2 are constants, to obtain the most central result in interpolation theory.

Corollary 1.6.11. *Let $A_0 \hookrightarrow A_1$ and $B_0 \hookrightarrow B_1$ be Banach spaces. Let $T\colon A_1 \to B_1$ be a bounded linear operator such that T is bounded from A_0 to B_0. Then for every $\theta \in (0,1)$ and $p \in [1,\infty)$ the operator T is bounded from $(A_0, A_1)_{\theta,p}$ to $(B_0, B_1)_{\theta,p}$.*

Proof. Since $\alpha = \beta = 1$, we have $\eta = \theta$ and $q = p$. Hence T maps $(A_0, A_1)_{\theta,p}$ into $(B_0, B_1)_{\theta,p}$. Then, by the closed graph theorem, T is a bounded operator between these spaces. □

Next, we apply Theorem 1.6.9 to a scale of products of Lebesgue spaces. This is not a straight-forward corollary because the real interpolation space of two Lebesgue spaces is not a Lebesgue space if the interpolation parameters do not fit with the exponents of the Lebesgue spaces. Thus it is not possible to go in one step from the product of Lebesgue spaces to the product in the target space. Fortunately, however, the gap to the desired Lebesgue space is not large. So if we introduce some separation distance by using a parameter $\varepsilon > 0$, we can go back to Lebesgue spaces.

If we used the complex interpolation method instead, we would not have such a problem, and in fact this approach works flawlessly for linear problems. For nonlinear interpolation, on the other hand, the complex interpolation theory is not as powerful as the real one, to the effect that no complex variant of Theorem 1.6.9 is known that would apply to the situation in Section 3.2.

Theorem 1.6.12. *Let (Ω, μ) be a finite measure space, $N \in \mathbb{N}$, $\theta \in (0,1)$, $\alpha \in (0,1]$, $\beta \in (0,\infty)$, and let $p_0 \geq p_1$, $q_0 \geq q_1$, and $r_0 \geq r_1$ be real numbers in $[1,\infty)$, and let $s_0 \geq s_1$ and $t_0 \geq t_1$ be numbers in $[1,\infty]$. As a shorthand, we write*

$$A_{p,q,r} := L^p(\Omega) \oplus L^q(\Omega)^N \oplus L^r(\partial\Omega) \quad \text{and} \quad B_{s,t} := L^s(\Omega) \oplus L^t(\partial\Omega)$$

and equip these spaces with natural norms. Let T be an operator that maps A_{p_1,q_1,r_1} into B_{s_1,t_1} such that $A_{p_0+\varepsilon,q_0+\varepsilon,r_0+\varepsilon}$ is continuously mapped into B_{s_0,t_0} for every $\varepsilon > 0$. Assume moreover that for every $\varepsilon > 0$ there exist locally bounded functions c_0 and c_1 such that

$$\|Tf - Tg\|_{B_{s_1,t_1}} \leq c_0(\|f\|_{A_{p_1,q_1,r_1}}, \|g\|_{A_{p_1,q_1,r_1}})\, \|f-g\|^{\alpha}_{A_{p_1,q_1,r_1}}$$

for all f and g in A_{p_1,q_1,r_1} and

$$\|Tf\|_{B_{s_0,t_0}} \leq c_1(\|f\|_{A_{p_1,q_1,r_1}})\, (\|f\|^{\beta}_{A_{p_0+\varepsilon,q_0+\varepsilon,r_0+\varepsilon}} + 1)$$

for all $f \in A_{p_0+\varepsilon,q_0+\varepsilon,r_0+\varepsilon}$. We define $\eta \in (0,1)$ by

$$\frac{1-\eta}{\eta} = \frac{1-\theta}{\theta}\frac{\alpha}{\beta}$$

and let

$$\frac{1}{p_\theta} := \frac{\theta}{p_1} + \frac{1-\theta}{p_0}, \qquad \frac{1}{q_\theta} := \frac{\theta}{q_1} + \frac{1-\theta}{q_0}, \qquad \frac{1}{r_\theta} := \frac{\theta}{r_1} + \frac{1-\theta}{r_0},$$
$$\frac{1}{s_\eta} := \frac{\eta}{s_1} + \frac{1-\eta}{s_0}, \qquad \frac{1}{t_\eta} := \frac{\eta}{t_1} + \frac{1-\eta}{t_0}.$$

Then for every $\varepsilon > 0$, the operator T maps $A_{p_\theta+\varepsilon,q_\theta+\varepsilon,r_\theta+\varepsilon}$ into B_{s_η,t_η}.

Proof. For simplicity we assume that $p_0 > p_1$, $q_0 > q_1$, $r_0 > r_1$, $s_0 > s_1$, and $t_0 > t_1$. In fact, otherwise there is no need for interpolation in the corresponding factor, so we can simply omit it in the considerations that follow.

Let $\varepsilon > 0$ be arbitrary. Define for $\delta > 0$ and $\theta' \in [0,1]$ the numbers $p(\theta',\delta)$, $q(\theta',\delta)$ and $r(\theta',\delta)$ by

$$\frac{1}{p(\theta',\delta)} := \frac{\theta'}{p_1} + \frac{1-\theta'}{p_0+\delta}, \qquad \frac{1}{q(\theta',\delta)} := \frac{\theta'}{q_1} + \frac{1-\theta'}{q_0+\delta}, \qquad \frac{1}{r(\theta',\delta)} := \frac{\theta'}{r_1} + \frac{1-\theta'}{r_0+\delta}.$$

Since $p(\theta,0) = p_\theta$, $q(\theta,0) = q_\theta$, and $r(\theta,0) = r_\theta$, by continuity there exist $\theta_0 < \theta$ and $\delta_0 > 0$ such that $p(\theta_0,\delta_0) < p_\theta + \varepsilon$, $q(\theta_0,\delta_0) < q_\theta + \varepsilon$, and $r(\theta_0,\delta_0) < r_\theta + \varepsilon$.

Let $w \in [1,\infty)$ be such that $w \geq \beta$, and fix $\theta_1 \in (\theta_0,\theta)$. Then, by Theorem 1.6.8 and Lemma 1.6.5,

$$\begin{aligned} L^{p_\theta+\varepsilon}(\Omega) \hookrightarrow L^{p(\theta_0,\delta_0)}(\Omega) &= \left(L^{p_0+\delta_0}(\Omega), L^{p_1}(\Omega)\right)_{\theta_0,p(\theta_0,\delta_0)} \\ &\hookrightarrow \left(L^{p_0+\delta_0}(\Omega), L^{p_1}(\Omega)\right)_{\theta_1,w}, \end{aligned}$$

and similar embeddings hold for q_θ and r_θ. Thus, by Lemma 1.6.6,

$$A_{p_\theta+\varepsilon,q_\theta+\varepsilon,r_\theta+\varepsilon} \hookrightarrow \mathcal{A}_{\theta_1} := \left(A_{p_0+\delta_0,q_0+\delta_0,r_0+\delta_0}, A_{p_1,q_1,r_1}\right)_{\theta_1,w}. \tag{1.6.10}$$

Let $\eta_1 \in (0,1)$ be given by

$$\frac{1-\eta_1}{\eta_1} = \frac{1-\theta_1}{\theta_1}\frac{\alpha}{\beta}$$

and define

$$\mathcal{B}_{\eta_1} := \left(B_{s_0,t_0}, B_{s_1,t_1}\right)_{\eta_1,\tilde{w}},$$

where $\tilde{w} = \frac{w}{(1-\eta_1)\beta+\eta_1\alpha}$. Since $\eta_1 < \eta$, by Lemma 1.6.5 and Theorem 1.6.8 we obtain that

$$\left(L^{s_0}(\Omega), L^{s_1}(\Omega)\right)_{\eta_1,\tilde{w}} \hookrightarrow \left(L^{s_0}(\Omega), L^{s_1}(\Omega)\right)_{\eta,s_\eta} = L^{s_\eta}(\Omega).$$

A similar embedding holds for t_η. Thus, by Lemma 1.6.6,

$$\mathcal{B}_{\eta_1} \hookrightarrow B_{s_\eta,t_\eta}. \tag{1.6.11}$$

Now, since T maps $\mathcal{A}_{\theta_1}$ into $\mathcal{B}_{\eta_1}$ by Theorem 1.6.9, the theorem follows from the inclusions (1.6.10) and (1.6.11). □

1.7 Linear semigroups

In this section we collect some facts from linear semigroup theory that will be needed in Sections 4.1 and 4.2 to show that certain problems are well-posed in the space $\mathrm{C}(\overline{\Omega})$. Much more information on the whole topic can be found in the books by Engel and Nagel [EN00], Ouhabaz [Ouh05] and Arendt, Batty, Hieber, and Neubrander [ABHN01], the latter being the main reference for this collection of results.

Definition 1.7.1 (C_0-semigroup, [ABHN01, Definition 3.1.8]). Let X be a Banach space. A function $T\colon [0,\infty) \to \mathscr{L}(X)$ is called a *(linear)* C_0*-semigroup on* X if

(i) $T(0) = I$;

(ii) $T(t+s) = T(t)T(s)$ for all $s, t > 0$;

(iii) $t \mapsto T(t)x$ is continuous on $[0,\infty)$ for every $x \in X$.

In this case, there exist $M \geq 1$ and $\omega \in \mathbb{R}$ such that

$$\|T(t)\| \leq M\,\mathrm{e}^{\omega t} \text{ for all } t \geq 0.$$

Moreover, there exists a unique linear operator A on X such that $\lambda - A$ is invertible for $\lambda > \omega$ and

$$R(\lambda, A)x = \int_0^\infty \mathrm{e}^{-\lambda t}\, T(t)x \,\mathrm{d}t \tag{1.7.1}$$

for all $x \in X$ and $\lambda > \omega$. We call A the *(infinitesimal) generator* of T.

We quote a famous theorem characterizing generators of C_0-semigroups.

Theorem 1.7.2 (Hille-Yoshida, [ABHN01, Theorem 3.3.4 and Corollary 3.3.6]). *Let X be a Banach space, and let A be a densely defined operator on X. Assume that there exist constants $M \geq 1$ and $\omega \in \mathbb{R}$ such that $\lambda - A$ is invertible for all $\lambda > \omega$ with*

$$\|R(\lambda, A)^{(n)}\| \leq \frac{M n!}{(\lambda - \omega)^{n+1}}, \tag{1.7.2}$$

where $R(\lambda, A)^{(n)}$ denotes the nth derivative of the analytic function $\lambda \mapsto R(\lambda, A)$. Then A generates a C_0-semigroup $(T(t))_{t\geq 0}$ on X. Moreover,

$$T(t)u = \lim_{n\to\infty} \left(I - \tfrac{t}{n}A\right)^{-n} u \tag{1.7.3}$$

for all $t > 0$ and $u \in X$.

Remark 1.7.3. On reflexive spaces, an operator satisfying (1.7.2) is always densely defined. In fact, it already suffices that there exists a constant M' such that

$$\|\lambda R(\lambda, A)\| \leq M'$$

for all sufficiently large λ.

To see this, let $x \in X$ be arbitrary and define $x_\lambda := R(\lambda, A)x \in D(A)$. Then λx_λ remains bounded as λ tends to infinity. Hence the exists a sequence (λ_n) such that $\lambda_n x_{\lambda_n} \rightharpoonup z$ for some $z \in X$. Since $\lambda_n x_{\lambda_n} - Ax_{\lambda_n} = x$ and $x_{\lambda_n} \to 0$, the (weak) closedness of the graph of A implies that $z = x$. Thus x is in the (weak) closure of $D(A)$.

In Chapter 4 we will be interested in solutions of parabolic problems. It is well understood that the right notion for solutions is the following.

Definition 1.7.4 (Mild solution). Let X be a Banach space, $x \in X$, and let A be a densely defined linear operator on X. We say that $u\colon [0, \infty) \to X$ is a *mild solution* of the *abstract Cauchy problem*

$$\begin{cases} \dot{u}(t) = Au(t), & t \geq 0, \\ u(0) = x \end{cases} \tag{1.7.4}$$

if u is continuous, $\int_0^t u(s)\, \mathrm{d}s \in D(A)$ for all $t \geq 0$, and

$$u(t) = x + A \int_0^t u(s) \mathrm{d}s$$

for all $t \geq 0$.

The relation between semigroups and abstract Cauchy problems can be put into a very elegant form. This relation allows us to study abstract Cauchy problems via semigroup theory.

Theorem 1.7.5 ([ABHN01, Theorem 3.1.12]). *Let X be a Banach space and A be a closed operator on X. Then A generates a C_0-semigroup $(T(t))_{t\geq 0}$ if and only if for all $x \in X$ problem* (1.7.4) *has a unique mild solution. In this case, $u(t) := T(t)x$ is the mild solution of* (1.7.4).

On Hilbert spaces, there is a particularly convenient way to show that certain operators generate C_0-semigroups. In fact, if the operator is defined via a form, then generation is automatic, and various properties of the semigroup can be expressed in terms of the form.

Definition 1.7.6. Let V and H be Hilbert spaces, let V be densely embedded into H, and let $a\colon V \times V \to \mathbb{R}$ be a form. Then we call A given by

$$\begin{aligned} D(A) &:= \big\{u \in V : \exists f \in H \text{ such that } a(u, v) = (f \mid v)_H \text{ for all } v \in V\big\}, \\ Au &:= f \end{aligned}$$

the operator that is *associated with a*.

Theorem 1.7.7 ([Ouh05, Theorems 1.45 and 1.52 and Proposition 2.5]). *Let V and H be Hilbert spaces, and let $a\colon V \times V \to \mathbb{R}$ be a continuous, coercive form with associated operator A. Then $-A$ generates a C_0-semigroup $(T(t))_{t\geq 0}$ on H consisting of contraction operators. Moreover, $T(t)u \in D(A)$ for all $t > 0$ and $u \in H$ and $t \mapsto T(t)u$ is differentiable on $(0,\infty)$ with $\frac{\mathrm{d}}{\mathrm{d}t}T(t)u = -AT(t)u$ for all $t > 0$ and $u \in H$.*

Remark 1.7.8. We can relax the condition that a is coercive to the requirement that a is H-elliptic, i.e., that a^ω given by $a^\omega(u,v) := a(u,v) + \omega\,(u \mid v)_H$ is coercive for some $\omega \in \mathbb{R}$, and can still deduce that $-A$ generates a C_0-semigroup. In fact, a^ω is associated with the operator $A^\omega := A + \omega$ and we know that $-A^\omega$ generates a C_0-semigroup $(T(t))_{t\geq 0}$. It is now easy to deduce from the definitions that then $-A$ generates the C_0-semigroup $(\mathrm{e}^{\omega t}T(t))_{t\geq 0}$.

Theorem 1.7.9 ([Ouh05, Theorem 2.2]). *Let V and H be Hilbert spaces, and let $a\colon V \times V \to \mathbb{R}$ be a continuous, coercive form. Let A be the associated operator and T be the C_0-semigroup generated by $-A$ on H. Let C be a closed, convex subset of H, and let P be the orthogonal projection of H onto C. Then the following are equivalent:*

(i) $T(t)C \subset C$ *for all* $t \geq 0$*;*

(ii) $PV \subset V$ *and* $a(Pu, u - Pu) \geq 0$ *for all* $u \in V$.

(iii) $PV \subset V$ *and* $a(u, u - Pu) \geq 0$ *for all* $u \in V$.

Corollary 1.7.10 ([Ouh05, Corollary 2.17]). *Under the same assumptions as in Theorem 1.7.9, let $H = L^2(\Omega)$ for some finite measure space (Ω,μ). Then the semigroup $(T(t))_{t\geq 0}$ satisfies*

$$\|T(t)u\|_{L^\infty(\Omega)} \leq \|u\|_{L^\infty(\Omega)} \text{ for all } u \in L^\infty(\Omega) \text{ and all } t \geq 0$$

if and only if

$$(|u| \wedge 1)\operatorname{sgn}(u) \in V \text{ and } a\big((|u| \wedge 1)\operatorname{sgn}(u), (|u|-1)^+\operatorname{sgn}(u)\big) \geq 0 \text{ for all } u \in V. \quad (1.7.5)$$

Here, $|u| \wedge 1$ denotes the pointwise minimum of the functions $|u|$ and 1.

1.8 Nonlinear semigroups and subdifferentials

We collect some results about nonlinear semigroups and give special emphasis to semigroups arising from subdifferentials, which have several nice properties. In particular, for such semigroups invariance criteria exist that are similar to Theorem 1.7.9 in the linear case. Standard references on nonlinear semigroup theory include the books by Brézis [Bré73] and Bénilan, Crandall, and Pazy [BCP]. This summary is mainly based on Showalter's book [Sho97].

Definition 1.8.1 (Nonlinear C_0-semigroup)**.** Let X be a Banach space and C be a closed, convex subset of X. A family $(T(t))_{t\geq 0}$ of (nonlinear) operators from C to C is called a *nonlinear contraction* C_0*-semigroup on* C if

(i) $T(0) = I$;

(ii) $T(t+s) = T(t)T(s)$ for all $s, t > 0$;

(iii) $\|T(t)x - T(t)y\| \leq \|x - y\|$ for all $x, y \in C$ and $t > 0$;

(iv) $t \mapsto T(t)x$ is continuous on $[0, \infty)$ for every $x \in X$.

For general Banach spaces, in particular Banach spaces whose dual space is not uniformly convex, the theory of generators of nonlinear semigroups is rather subtle. There is no such universal characterization as Theorem 1.7.2 in the nonlinear case, but see [BCP, Theorem 18.3] for the Hilbert space setting. For our purposes it suffices to associate a nonlinear semigroup to an m-accretive operator, which is the family of solution operators for the evolution problem. It turns out that this definition is consistent with the linear theory, compare (1.8.2) and (1.7.3).

Definition 1.8.2 (m-accretive operator)**.** Let X be a Banach space and A be an operator from $D(A) \subset X$ to X. We call A *accretive* if

$$\|x_1 - x_2\| \leq \|(x_1 + \alpha A x_1) - (x_2 + \alpha A x_2)\|$$

holds for all x_1 and x_2 in $D(A)$ and all $\alpha > 0$. We call A *m-accretive* if in addition $I + \alpha A \colon D(A) \to X$ is surjective for some $\alpha > 0$.

Remark 1.8.3 ([Sho97, Corollary IV.1.3])**.** In a Hilbert space H, an operator A is accretive if and only if

$$(Au - Av \mid u - v) \geq 0 \text{ for all } u, v \in D(A),$$

compare also Definition 1.5.1.

Lemma 1.8.4 ([Sho97, Lemma IV.7.1])**.** *If A is an m-accretive operator on a Banach space X, then $I + \alpha A$ is surjective for every $\alpha > 0$.*

Definition 1.8.5 (Generated semigroup)**.** Let X be a Banach space and let A be an m-accretive mapping on X. Given $\varepsilon > 0$ and an interval $[a, b] \subset \mathbb{R}$, we say that a step function

$$v = v_0 \mathbb{1}_{\{t_0\}} + \sum_{i=1}^{N} v_i \mathbb{1}_{(t_{i-1}, t_i]}, \quad (a = t_0 < t_1 < \cdots < t_N = b)$$

is an ε*-solution of*

$$u'(t) + Au(t) = 0 \tag{1.8.1}$$

on $[a, b]$ if

$$t_i - t_{i-1} < \varepsilon \text{ for all } i = 1, \ldots, N$$

and

$$\sum_{i=1}^{N} \left\| v_i - v_{i-1} + (t_i - t_{i-1})A(v_i) \right\| < \varepsilon.$$

We say that $u \in \mathrm{C}([0,\infty); X)$ is a *mild solution of* (1.8.1) if for every $T > 0$ and every $\varepsilon > 0$ there exists an ε-solution v of (1.8.1) on $[0,T]$ such that $\|u(t) - v(t)\| \leq \varepsilon$ for all $t \in [0,T]$. Finally, given a nonlinear contraction C_0-semigroup $(T(t))_{t\geq 0}$ on $\overline{D(A)}$, we say that A *generates* $(T(t))_{t\geq 0}$ if for every $u_0 \in \overline{D(A)}$ the mapping $t \mapsto T(t)u_0$ is a mild solution of (1.8.1).

The following theorem is the main generation result for nonlinear contraction C_0-semigroups in arbitrary Banach spaces. For the proof, confer [Sho97, Theorems IV.8.1 and IV.8.2] or [BCP, Proposition 3.7 and Theorem 3.4].

Theorem 1.8.6 (Crandall-Liggett). *Let X be a Banach space, and let A be an m-accretive operator from $D(A) \subset X$ to X. Then for every $u_0 \in \overline{D(A)}$ there exists a unique mild solution u of* (1.8.1) *that satisfies $u(0) = u_0$. Moreover,*

$$u(t) = \lim_{m\to\infty} \left(I + \tfrac{t}{m}A\right)^{-m} u_0 \tag{1.8.2}$$

uniformly in $t \in [0,T]$ for every $T > 0$. Write $T(t)u_0 := u(t)$ for $u_0 \in \overline{D(A)}$ and $t \geq 0$, where u denotes the unique mild solution of (1.8.1) *with $u(0) = u_0$. Then*

$$\|T(t)u_0 - T(t)v_0\| \leq \|u_0 - v_0\|$$

for all u_0 and v_0 in $\overline{D(A)}$ and $t \geq 0$. In other words, $(T(t))_{t\geq 0}$ is a nonlinear contraction C_0-semigroup on $\overline{D(A)}$ with generator A.

In Hilbert spaces, subdifferentials of convex energy functionals constitute examples of m-accretive operators. Because this is a simple way to construct examples of m-accretive operators, this approach to nonlinear semigroups is very convenient whenever it applies. An additional motivation for us to introduce subdifferentials is the fact that invariance properties of nonlinear semigroups can be expressed in terms of the energy functional, compare Theorem 1.8.11.

Definition 1.8.7 (Subdifferential). Let V be a Banach space. A (nonlinear) functional $\varphi\colon V \to (-\infty,\infty]$ is called *proper* if its *domain*

$$\mathrm{dom}(\varphi) := \{u \in V : \varphi(u) < \infty\}$$

is non-empty. It is called *convex* if

$$\varphi(\lambda u + (1-\lambda)v) \leq \lambda\varphi(u) + (1-\lambda)\varphi(v)$$

holds for all u and v in V and all $\lambda \in [0,1]$. It is called *lower semicontinuous* if the set $\varphi^{-1}((-\infty,c])$ is closed in V for every $c \in \mathbb{R}$, or, which is the same by [Sho97,

Proposition IV.1.3],

$$\varphi(u_0) \le \liminf_{u \to u_0} \varphi(u)$$

for all $u_0 \in \operatorname{dom}(\varphi)$. Given a proper, convex, lower semicontinuous functional $\varphi\colon V \to (-\infty, \infty]$ we define

$$\partial\varphi(u) := \begin{cases} \emptyset, & u \notin \operatorname{dom}(\varphi), \\ \{f \in V' : \langle f, v-u\rangle_{V',V} \le \varphi(v) - \varphi(u)\ \forall v \in V\}, & u \in \operatorname{dom}(\varphi) \end{cases}$$

and call the *set-valued* (or *multi-valued*) mapping $\partial\varphi$ the *subdifferential* of φ. Its *domain* is

$$D(\partial\varphi) := \{u \in V : \partial\varphi(u) \neq \emptyset\}.$$

If $\partial\varphi$ is *single-valued*, i.e., for every $u \in V$ the set $\partial\varphi(u)$ contains at most one element, then we frequently regard $\partial\varphi$ as an operator from $D(\partial\varphi)$ to V'. We say that φ is *Gâteaux-differentiable* at an interior point u of $\operatorname{dom}(\varphi)$ if there exists $f \in V'$ such that

$$\langle f, v\rangle_{V',V} = \lim_{t \to 0} \frac{\varphi(u+tv) - \varphi(u)}{t} \quad \text{for every } v \in V,$$

and in this case we write $\varphi'(u) := f$. If $V = H$ is a Hilbert space, we frequently identify it with its dual via the Riesz-Fréchet representation theorem and regard $\partial\varphi(u)$ as a subset of H and $\partial\varphi$ as an operator on H if $\partial\varphi$ is single-valued.

In general, we have $D(\partial\varphi) \subsetneq \operatorname{dom}(\varphi)$. Still, the following lemma is a simple criterion that allows us to check that $\partial\varphi$ is densely defined.

Lemma 1.8.8 ([Sho97, Corollary IV.1.4]). *Let $\varphi\colon H \to (-\infty, \infty]$ be proper, convex, and lower semicontinuous on a Hilbert space H. Then $\operatorname{dom}(\varphi) \subset \overline{D(\partial\varphi)}$.*

It will be convenient for the examples to consider the functional first on a smaller Banach space on which it is nicely behaved and then to extend it to the Hilbert space under consideration. The following lemma allows us to do so.

Lemma 1.8.9. *Let V be a reflexive Banach space and let H be a Hilbert space into which V is continuously embedded. We regard V as a subspace of H. Let $\varphi\colon V \to (-\infty, \infty]$ be a proper, convex, lower semicontinuous functional. Assume that the sets*

$$C_{a,b} := \{u \in V : \varphi(u) \le a \text{ and } \|u\|_H \le b\}$$

are bounded in V for every $a, b \ge 0$. Then

$$\tilde\varphi(u) := \begin{cases} \varphi(u), & u \in V, \\ \infty, & u \in H \setminus V \end{cases}$$

defines a proper, convex, lower semicontinuous functional $\tilde\varphi\colon H \to (-\infty, \infty]$. Moreover, *if φ is Gâteaux-differentiable at an interior point u of* $\operatorname{dom}(\varphi)$*, then*

$$\partial\tilde{\varphi}(u) = \left\{ f \in H = H' : (f \mid v)_H = \langle \varphi'(u), v \rangle_{V',V} \ \forall v \in V \right\} \tag{1.8.3}$$

for all $u \in V$. Thus if in addition V is dense in H, then $\partial\tilde{\varphi}(u)$ consists of at most one element.

Proof. It is clear that $\tilde{\varphi}$ is proper and convex. To see that $\tilde{\varphi}$ is lower semicontinuous, let (u_n) be a sequence in H converging to $u \in H$. In particular there exists $b \geq 0$ such that $\|u_n\|_H \leq b$ for all $n \in \mathbb{N}$. We have to show that

$$\liminf_{n\to\infty} \tilde{\varphi}(u_n) \geq \tilde{\varphi}(u).$$

Passing to a subsequence, we can assume without loss of generality that $(\tilde{\varphi}(u_n))$ converges to some finite value. Let $a > \lim_{n\to\infty} \tilde{\varphi}(u_n)$ be arbitrary. Then the sequence (u_n) is eventually contained in $C_{a,b}$. Thus, by assumption, (u_n) is bounded in V, and again passing to a subsequence we may assume that there exists $v \in V$ such that $u_n \rightharpoonup v$ in V. Since φ is lower semicontinuous and convex, the set

$$L_a := \{w \in V : \varphi(w) < a\} \subset C_{a,b}$$

is closed and convex in V. Hence $v \in L_a$ by the Hahn-Banach theorem, see [Lax02, Theorem 10.6]. Moreover, $u_n \rightharpoonup v$ also in H since $V \hookrightarrow H$, so $u = v \in V$ and $\varphi(u) \leq a$. Since $a > \lim_{n\to\infty} \varphi(u_n)$ was arbitrary, this proves

$$\tilde{\varphi}(u) = \varphi(u) \leq \lim_{n\to\infty} \varphi(u_n) = \liminf_{n\to\infty} \tilde{\varphi}(u_n).$$

We have shown that $\tilde{\varphi}$ is lower semicontinuous.

For (1.8.3), assume that φ is Gâteaux-differentiable at an interior point u of $\operatorname{dom}(\varphi) \subset V$. Let $f \in \partial\tilde{\varphi}(u)$. Then by definition

$$(f \mid v - u)_H \leq \tilde{\varphi}(v) - \tilde{\varphi}(u)$$

for all v in H. Setting $v = u \pm tw$, $t > 0$, we obtain in particular that

$$\pm (f \mid w)_H \leq \pm \frac{\varphi(u + tw) - \varphi(u)}{t}$$

for all $t > 0$ and all $w \in V$. Passing to the limit as $t \to 0$ and exploiting the fact that the inequality holds with both signs, we deduce that

$$(f \mid w)_H = \langle \varphi'(u), w \rangle_{V',V}$$

for all $w \in V$ by definition of the Gâteaux derivative. This proves one of the inclusions in (1.8.3).

For the converse inclusion assume that $f \in H$ is such that

$$(f \mid v)_H = \langle \varphi'(u), v\rangle_{V',V}$$

for all $v \in V$. Since

$$\varphi(u + t(v-u)) = \varphi(tv + (1-t)u) \le t\varphi(v) + (1-t)\varphi(u) = \varphi(u) + t(\varphi(v) - \varphi(u))$$

for all $v \in V$ and $t \in [0,1]$ and thus

$$\frac{\varphi(u + t(v-u)) - \varphi(u)}{t} \le \varphi(v) - \varphi(u),$$

we obtain by taking the limit $t \to 0$ that

$$(f \mid v-u)_H = \langle \varphi'(u), v-u\rangle_{V',V} \le \varphi(v) - \varphi(u)$$

for all $v \in V$. Since the inequality is trivially true also if $v \notin V$, we obtain that $f \in \partial\tilde{\varphi}(u)$. □

In our examples in Section 4.3 all the subdifferentials are single-valued. Thus for notational simplicity we from now on always assume this as a condition, even though there is no mathematical need for this restriction. In fact, the whole theory presented below remains valid for multi-valued operators without significant changes.

As already stated, subdifferentials are particular instances of m-accretive operators in Hilbert spaces, see [Sho97, Proposition IV.1.5 and Corollary IV.1.3]. Thus by Theorem 1.8.6 every subdifferential generates a nonlinear contraction C_0-semigroup. In this setting the mild solutions of (1.8.1) have much nicer properties than in the general case; see [Sho97, Propositions IV.3.1 and IV.3.2] for a proof of the following theorem.

Theorem 1.8.10 (Kōmura-Kato and Brézis). *Let $\varphi\colon H \to (-\infty, \infty]$ be proper, convex, and lower semicontinuous on a Hilbert space H, and set $A := \partial\varphi$. For simplicity assume that A is single-valued. Then A generates a nonlinear contraction C_0-semigroup $(T(t))_{t\ge 0}$ on $\overline{D(A)}$. Moreover, for every $u_0 \in \overline{D(A)}$ the mild solution u of $u'(t) + Au(t) = 0$ satisfying $u(0) = u_0$, i.e., $u(t) := T(t)u_0$, is locally Lipschitz continuous on $(0,\infty)$, thus almost everywhere differentiable, maps into $D(A)$, and satisfies $u'(t) + Au(t) = 0$ almost everywhere.*

We are now in the position to state the announced invariance criterion for nonlinear C_0-semigroups arising from subdifferentials.

Theorem 1.8.11. *Let (Ω, μ) be a finite measure space. Let $\varphi\colon H \to (-\infty, \infty]$ be proper, convex, and lower semicontinuous on $H := L^2(\Omega)$. We denote by $(T(t))_{t\ge 0}$ the nonlinear contraction C_0-semigroup generated by $A := \partial\varphi$. For u and v in H and $\lambda > 0$ define*

$\varphi^{(2)}(u,v) := \varphi(u) + \varphi(v)$ and

$$P_\lambda(u,v) := \begin{cases} (u,v) & \text{on } \{|u-v| \leq \lambda\}, \\ (\frac{1}{2}(u+v-\lambda), \frac{1}{2}(u+v+\lambda)) & \text{on } \{u-v < -\lambda\}, \\ (\frac{1}{2}(u+v+\lambda), \frac{1}{2}(u+v-\lambda)) & \text{on } \{u-v > \lambda\}. \end{cases}$$

Then the following assertions are equivalent, where in the second and third assertion it may happen that both sides of the inequality are infinite.

(i) $\varphi^{(2)}(P_\lambda(u,v)) \leq \varphi^{(2)}(u,v)$ *for all u and v in H and $\lambda > 0$.*

(ii) $\|T(t)u - T(t)v\|_{L^\infty(\Omega)} \leq \|u-v\|_{L^\infty(\Omega)}$ *for all $t \geq 0$ and all u and v in H.*

(iii) $\|u_1 - u_2\|_{L^\infty(\Omega)} \leq \|(u_1 + \alpha A u_1) - (u_2 + \alpha A u_2)\|_{L^\infty(\Omega)}$ *for $\alpha > 0$ and $u_1, u_2 \in D(A)$.*

Proof. The equivalence of the first two assertions is the statement is [CG03, Theorem 3.6]. Moreover, this is equivalent to the fact that for every $\lambda > 0$ the closed, convex set

$$C_\lambda := \{(u,v) \in H \times H : \|u-v\|_{L^\infty(\Omega)} \leq \lambda\}$$

is invariant under the action of the semigroup $T^{(2)}(t)(u,v) := (T(t)u, T(t)v)$. The semigroup $(T^{(2)}(t))_{t\geq 0}$ has the generator $A^{(2)}$ acting as $A^{(2)}(u,v) = (Au, Av)$ on its domain $D(A^{(2)}) = D(A) \times D(A)$. Thus by [Bré73, Proposition 4.5] the first two assertions are equivalent to the condition that $(u,v) \mapsto ((I+\alpha A)^{-1}u, (I+\alpha A)^{-1}v)$ leaves C_λ invariant for all $\lambda > 0$ and $\alpha > 0$, i.e.,

$$\|(I+\alpha A)^{-1}v_1 - (I+\alpha A)^{-1}v_2\|_{L^\infty(\Omega)} \leq \|v_1 - v_2\|_{L^\infty(\Omega)}$$

for all $v_1, v_2 \in H$ and $\alpha > 0$. Since A is m-accretive, we can express v_1 and v_2 uniquely as $v_1 = u_1 + \alpha A u_1$ and $v_2 = u_2 + \alpha A u_2$ with $u_1, u_2 \in D(A)$, see Lemma 1.8.4, thus obtaining the equivalence to the third assertion. □

1.9 Differential equations

We will consider partial equations in divergence form on Lipschitz domains Ω in $\mathbb{R}^N$. Our main focus is on quasi-linear equations. However, to treat those it will help to have at hand some theory for the linear case as well.

In this section, we introduce the terminology that we use in the following chapters regarding weak solutions and boundary conditions. Moreover, for later application we quote interior regularity results for solutions of quasilinear equations.

The most general kind of differential equation that we consider is formally given by

$$-\operatorname{div} A(x,u,\nabla u) + a(x,u,\nabla u) = 0 \tag{1.9.1}$$

with measurable functions $A\colon \Omega \times \mathbb{R} \times \mathbb{R}^N \to \mathbb{R}^N$ and $a\colon \Omega \times \mathbb{R} \times \mathbb{R}^N \to \mathbb{R}$. We will add appropriate ellipticity, regularity, and growth assumptions later, depending on the situation. As a convention that turns out to be convenient for the calculations, the third argument p is always a row vector and A takes values in the space of column vectors.

Often it will be useful to write (1.9.1) in the inhomogeneous form

$$-\operatorname{div} A(x,u,\nabla u) + a(x,u,\nabla u) = f_0(x) - \sum_{j=1}^{N} D_j f_j(x) \tag{1.9.2}$$

with functions f_j $(j = 1, \ldots, N)$ in $L^1_{\text{loc}}(\Omega)$.

A function in $W^{1,1}_{\text{loc}}(\Omega)$ is called a *weak solution* of (1.9.1) or (1.9.2) if

$$\int_\Omega \nabla\eta\, A(x,u,\nabla u) + \int_\Omega \eta\, a(x,u,\nabla u) = 0$$

or

$$\int_\Omega \nabla\eta\, A(x,u,\nabla u) + \int_\Omega \eta\, a(x,u,\nabla u) = \int_\Omega \eta\, f_0 + \sum_{j=1}^{N} \int_\Omega D_j\eta\, f_j$$

for all test functions $\eta \in \mathrm{C}_c^\infty(\Omega)$, respectively. Here, it is implicit in the definition that the integrals exist. It is easy to justify this definition by integration by parts. In fact, if all involved functions are sufficiently smooth, then it is easy to see that u is a weak solution if and only if u satisfies (1.9.1) or (1.9.2), respectively, in a classical sense.

A particular instance of quasilinear equations are linear problems, i.e., equations of the form

$$-\sum_{j=1}^{N} D_j\Big(\sum_{i=1}^{N} a_{ij} D_i u + b_j u\Big) + \Big(\sum_{i=1}^{N} c_i D_i u + du\Big) = f_0 - \sum_{j=1}^{N} D_j f_j \tag{1.9.3}$$

with measurable functions a_{ij}, b_j, c_i, d, f_0, and f_j. Linear problems are of the form (1.9.1) with

$$A_j(x,u,\nabla u) = \sum_{i=1}^{N} a_{ij} D_i u + b_j u - f_j, \qquad a(x,u,\nabla u) = \sum_{i=1}^{N} c_i D_i u + du - f_0 \tag{1.9.4}$$

or, equivalently, of the form (1.9.2) with

$$A_j(x,u,\nabla u) = \sum_{i=1}^{N} a_{ij} D_i u + b_j u, \qquad a(x,u,\nabla u) = \sum_{i=1}^{N} c_i D_i u + du. \tag{1.9.5}$$

Consistently, we say that $u \in W^{1,1}_{\mathrm{loc}}(\Omega)$ is a weak solution of (1.9.3) if

$$\sum_{i,j=1}^{N} \int_\Omega a_{ij} D_i u\, D_j \eta + \sum_{j=1}^{N} \int_\Omega b_j u\, D_j \eta + \sum_{i=1}^{N} \int_\Omega c_i D_i u\, \eta + \int_\Omega d u \eta = \int_\Omega f_0 \eta + \sum_{j=1}^{N} \int_\Omega f_j\, D_j \eta$$

for all test functions $\eta \in \mathrm{C}_c^\infty(\Omega)$.

Typically there are many solutions to the problems (1.9.1) and (1.9.3) on a given domain $\Omega \subset \mathbb{R}^N$. Therefore one usually specifies in addition some boundary conditions in the hope to obtain a unique solution to the problem. We say that $u \in W^{1,1}(\Omega)$ is a weak solution of the (quasi-linear) *Robin problem*

$$\begin{cases} -\operatorname{div} A(x,u,\nabla u) + a(x,u,\nabla u) = f_0(x) - \sum_{j=1}^{N} D_j f_j(x) & \text{on } \Omega, \\ A(z,u,\nabla u) \cdot \nu(z) + h(z,u) = g(z) + \sum_{j=1}^{N} f_j(z) \nu_j(z) & \text{on } \partial\Omega, \end{cases} \tag{1.9.6}$$

where $\nu = (\nu_j)$ denotes the outer normal of Ω, if

$$\int_\Omega \nabla \eta\, A(x,u,\nabla u) + \int_\Omega \eta\, a(x,u,\nabla u) + \int_{\partial\Omega} \eta\, h(z,u) = \int_\Omega \eta\, f_0 + \int_\Omega D_j \eta\, f_j + \int_{\partial\Omega} \eta\, g$$

holds for all $\eta \in \mathrm{C}^\infty(\overline{\Omega})$. The Robin problem in the special case $h(z,u) \equiv 0$, i.e.,

$$\begin{cases} -\operatorname{div} A(x,u,\nabla u) + a(x,u,\nabla u) = f_0(x) - \sum_{j=1}^{N} D_j f_j(x) & \text{in } \Omega, \\ A(z,u,\nabla u) \cdot \nu(z) = g(z) + \sum_{j=1}^{N} f_j(z) \nu_j(z) & \text{on } \partial\Omega, \end{cases} \tag{1.9.7}$$

is also called *Neumann problem*. Again, this definition can be justified by integration by parts, or, more precisely, by the divergence theorem.

We will apply the following results about interior regularity to solutions of (1.9.1), which are all of the same kind as the celebrated theorem due to De Giorgi and Nash. Since the assumptions of these theorems will reappear on a regular basis, they receive their own numbers for easier reference We start with a quasi-linear version of the theorem.

Assumption 1.9.1. *Let $m \in (1,\infty)$ be fixed, and let the functions A and a be measurable. Assume that there exist positive constants ν and μ and functions ψ_k such that*

$$p\, A(x,u,p) \geq \nu |p|^m - \psi_1 |u|^m - \psi_2, \tag{1.9.8}$$

$$\left| A(x,u,p) \right| \leq \mu |p|^{m-1} + \psi_3 |u|^{m-1} + \psi_4, \tag{1.9.9}$$

$$\left| a(x,u,p) \right| \leq \psi_5 |p|^{m-1} + \psi_6 |u|^{m-1} + \psi_7 \tag{1.9.10}$$

for almost all $x \in \Omega$, all $p \in \mathbb{R}^N$ and all $u \in \mathbb{R}$. If $m \le N$ we assume that

$$\psi_1, \psi_2, \psi_6, \psi_7 \in L^{\frac{N}{m}+\varepsilon}(\Omega), \ \ \psi_3, \psi_4 \in L^{\frac{N}{m-1}+\varepsilon}(\Omega), \ \ \text{and } \psi_5 \in L^{N+\varepsilon}(\Omega)$$

for some $\varepsilon \in (0, m-1)$. If $m > N$, we assume that

$$\psi_1, \psi_2, \psi_6, \psi_7 \in L^1(\Omega), \ \ \psi_3, \psi_4 \in L^{\frac{m}{m-1}}(\Omega), \ \ \text{and } \psi_5 \in L^m(\Omega).$$

Theorem 1.9.2 ([Ser64, Sections 1, 4, and 5]). *Let $\Omega \subset \mathbb{R}^N$ be open, and let Ω' be relatively compact in Ω. Under Assumption 1.9.1, there exist $\alpha \in (0,1)$ and $c \ge 0$ not depending on ψ_2, ψ_4, and ψ_7 such that every weak solution $u \in W^{1,m}(\Omega)$ of (1.9.1) is in $C^{0,\alpha}(\Omega')$ and*

$$\|u\|_{C^{0,\alpha}(\Omega)} \le c\Big(\|u\|_{L^m(\Omega)} + \big(\|\psi_4\| + \|\psi_7\|\big)^{1/(m-1)} + \|\psi_2\|^{1/m}\Big) \tag{1.9.11}$$

Here, the norms of the functions ψ_2, ψ_4, and ψ_7 are taken in the spaces indicated in Assumption 1.9.1.

For easier reference, we quote a linear version of this result as well, even though it is merely a special case of Theorem 1.9.2.

Assumption 1.9.3. *Assume that $a_{ij} \in L^\infty(\Omega)$, $b_j, c_i \in L^q(\Omega)$, and $d \in L^{q/2}(\Omega)$ for some $q > N$ and that there exists $\nu > 0$ such that*

$$\sum_{i,j=1}^{N} a_{ij}\xi_i\xi_j \ge \nu|\xi|^2$$

for all $x \in \Omega$ and $\xi \in \mathbb{R}^N$.

Theorem 1.9.4 ([LU68, Theorem 3.14.1]). *Let $\Omega \subset \mathbb{R}^N$ be open, let Ω' be relatively compact in Ω, and let $q > N$. Under Assumption 1.9.3, there exist $\alpha \in (0,1)$ and constant $c \ge 0$ with the following property. If $f_0 \in L^{q/2}(\Omega)$ and $f_j \in L^q(\Omega)$, then every weak solution $u \in W^{1,2}(\Omega)$ of (1.9.3) is in $C^{0,\alpha}(\Omega')$ and satisfies*

$$\|u\|_{C^{0,\alpha}(\Omega')} \le c\Big(\|u\|_{L^2(\Omega)} + \|f_0\|_{L^{q/2}(\Omega)} + \|f_j\|_{L^q(\Omega)}\Big).$$

2 Elliptic equations with Neumann boundary conditions

In this chapter we investigate existence, uniqueness, regularity, and continuous dependence of the right hand side for solutions of Neumann problems of the form (1.9.7). The approaches to existence and continuous dependence are rather canonical, but the idea how to obtain regularity seems to be new, and it has certainly not been exploited to this extent until now.

By investigating the Neumann problem in detail, we lay the foundation for proving regularity results for Robin problems in Chapter 3. In fact, almost all of the theorems proved here will have applications in Chapter 3.

2.1 Hölder regularity

In this section we will encounter the surprising fact that the same conditions as for interior Hölder regularity allow us to prove Hölder continuity up to the boundary for the corresponding Neumann problem on a bounded Lipschitz domain Ω. The strategy of the proof consists in extending u to a neighborhood of $\partial\Omega$, showing that this extension again solves an elliptic equation, and then applying Theorem 1.9.2.

Let $\Omega \subset \mathbb{R}^N$ be a bounded Lipschitz domain and $z \in \partial\Omega$. Then, by definition, we can choose an orthogonal matrix $\mathcal{O}$, a radius $r > 0$, and a Lipschitz continuous function ψ on $\mathbb{R}^{N-1}$ with $\psi(0) = 0$ such that for

$$G := \Big\{(y, \psi(y) + s) : y \in B(0, r) \subset \mathbb{R}^{N-1}, s \in (-r, r)\Big\}$$

we have

$$\mathcal{O}(\Omega - z) \cap G = \Big\{(y, \psi(y) + s) : y \in B(0, r) \subset \mathbb{R}^{N-1}, s \in (0, r)\Big\}.$$

In this situation, we define $T(y, s) := (y, \psi(y) + s)$ for $y \in \mathbb{R}^{N-1}$ and $s \in \mathbb{R}$. Moreover, we define the reflection $R\colon G \to G$ at the graph of ψ by $R(T(y, s)) := T(y, -s)$ and let $S\colon H \to H$, where $H := \mathcal{O}^{-1}G + z$, be the corresponding reflection at $\partial\Omega$, i.e.,

$$Sx := \mathcal{O}^{-1}R(\mathcal{O}(x - z)) + z.$$

In what follows, we need several properties of S.

Lemma 2.1.1. *The function S, defined on the neighborhood H of $z \in \partial\Omega$, has the following properties:*

(i) S is bi-Lipschitz, hence differentiable almost everywhere;

(ii) $S(Sx) = x$, $\det S'(x) \equiv -1$, $S'(x)^{-1} = S'(x)$, and $S'(Sx) = S'(x)$;

(iii) there exists $\alpha > 0$ such that $|\xi^T S'(x)| \geq \alpha|\xi|$ for all $\xi \in \mathbb{R}^N$.

Proof. For (i), note that T is a bi-Lipschitz mapping from $B(0,r) \times (-r,r)$ to G with derivative

$$T'(y,s) = \begin{pmatrix} I & 0 \\ \nabla\psi(y) & 1 \end{pmatrix}$$

almost everywhere. Thus S is bi-Lipschitz being a composition of bi-Lipschitz functions.

The fact $S(Sx) = x$ follows from $R(R(x)) = x$. By the chain rule, i.e., Theorem 1.2.5, applied to the defining equation of R, we have

$$R'(T(y,s))T'(y,s) = T'(y,s)\begin{pmatrix} I & 0 \\ 0 & -1 \end{pmatrix}$$

almost everywhere. Since $T'(y,s)$ is invertible with inverse

$$T'(y,s)^{-1} = \begin{pmatrix} I & 0 \\ -\nabla\psi(y) & 1 \end{pmatrix},$$

we obtain that

$$R'(T(y,s)) = T'(y,-s)\begin{pmatrix} I & 0 \\ 0 & -1 \end{pmatrix}T'(y,s)^{-1} = \begin{pmatrix} I & 0 \\ 2\nabla\psi(y) & -1 \end{pmatrix}$$

almost everywhere. Hence, by the chain rule,

$$S'(x) = \mathcal{O}^{-1}R'(\mathcal{O}(x-z))\mathcal{O} = \mathcal{O}^{-1}\begin{pmatrix} I & 0 \\ 2\nabla\psi(y) & -1 \end{pmatrix}\mathcal{O},$$

where y and s are such that $T(y,s) = \mathcal{O}(x-z)$. From this we obtain the remaining claims in part (ii).

For the proof of (iii), let $\xi \in \mathbb{R}^N$ be arbitrary and define $\zeta \in \mathbb{R}^{N-1}$ and $\omega \in \mathbb{R}$ by $(\zeta^T, \omega) = \xi^T\mathcal{O}^{-1}$. Let L denote the Lipschitz constant of ψ. Then $|\nabla\psi| \leq L$ almost everywhere. Pick $\delta \in (0,1)$ so small that $\frac{4\delta L^2}{1-\delta} \leq 1-\delta$. Then, since $\mathcal{O}$ is an isometry and by Young's inequality,

$$\begin{aligned} |\xi^T S'(x)|^2 &= \left|\begin{pmatrix} \zeta^T + 2\omega\nabla\psi(y) \\ -\omega \end{pmatrix}\right|^2 = |\zeta|^2 + 4\omega\nabla\psi\zeta + 4\omega^2|\nabla\psi|^2 + \omega^2 \\ &\geq |\zeta|^2 - (1-\delta)|\zeta|^2 - \frac{4}{1-\delta}\omega^2|\nabla\psi|^2 + 4\omega^2|\nabla\psi|^2 + \omega^2 \end{aligned}$$

$$\geq \delta|\zeta|^2 + \left(1 - \frac{4\delta}{1-\delta}|\nabla\psi|^2\right)\omega^2 \geq \delta\left(|\zeta|^2 + \omega^2\right) = \delta|\xi|^2.$$

Thus (iii) holds for $\alpha = \delta^{1/2}$. □

In what follows, we write U for $H \cap \Omega$ and V for $S(U) = H \setminus \overline{\Omega}$. For a function w on $D \subset H$, we define w^* by $w^*(x) := w(Sx)$ on $S(D)$. For a function w on U, we define $\tilde{w}$ almost everywhere on H by

$$\tilde{w}(x) := \begin{cases} w(x), & x \in U, \\ w^*(x) = w(Sx), & x \in V. \end{cases}$$

In the following it will not matter that $\tilde{w}$ is not defined on the Lebesgue null set $\partial\Omega \cap H$. In other words, one may define the function arbitrarily on this set.

Lemma 2.1.2. *Let $1 \leq m \leq \infty$.*

(i) If w is in $W^{1,m}(D)$, then w^* is in $W^{1,m}(S(D))$, and $\nabla w^*(x) = \nabla w(Sx)S'(x)$ almost everywhere.

(ii) If w is in $W^{1,m}(U)$, then $w|_{\partial U} = w^*|_{\partial V}$ on $\partial\Omega \cap H$.

(iii) If w is in $W^{1,m}(U)$, then $\tilde{w}$ is in $W^{1,m}(H)$ and $\nabla\tilde{w} = \nabla w\, \mathbb{1}_U + \nabla w^*\, \mathbb{1}_V$.

(iv) The extension operator $w \mapsto \tilde{w}$ is a bounded linear operator from $L^m(U)$ to $L^m(H)$.

Proof. Assertion (i) follows from Theorem 1.2.5. Assertion (ii) is obvious by the definition of the trace operator if w is in addition continuous up to the boundary. By continuity of the trace operator and since $C(\overline{\Omega}) \cap W^{1,m}(\Omega)$ is dense in $W^{1,m}(\Omega)$, (ii) follows by approximation.

Let $\varphi \in C_c^\infty(H)$. Remark 1.3.8 shows that

$$\int_H \tilde{w}\, D_i\varphi = \int_{\partial U} w\varphi\, \nu_i - \int_U D_i w\, \varphi + \int_{\partial V} w^*\varphi\, \nu_i - \int_V D_i w^*\, \varphi.$$

The boundary integrals cancel due to (ii) since the outer normals of U and V have opposite orientations on $\partial\Omega \cap H$. This proves (iii). Assertion (iv) follows from Theorem 1.2.3. □

Let A and a be fixed functions as in Assumption 1.9.1. We introduce on H the coefficients

$$\hat{A}(x,u,p) := \begin{cases} A(x,u,p), & x \in U, \\ S'(x)A(Sx,u,pS'(x)), & x \in V \end{cases}$$

and

$$\hat{a}(x,u,p) := \begin{cases} a(x,u,p), & x \in U, \\ a(Sx,u,pS'(x)), & x \in V \end{cases}$$

and show that these coefficients inherit the same regularity.

Lemma 2.1.3. *Let $L \geq 1$ be an upper bound for $|S'(x)|$, and let $\alpha \in (0,1)$ be as in part* (iii) *of Lemma 2.1.1. Then the coefficients $\hat{A}$ and $\hat{a}$ satisfy Assumption 1.9.1 for the parameters*

$$\begin{aligned} &\hat{\nu} := \alpha^m \nu, && \hat{\mu} := L^m \mu, && && \\ &\hat{\psi}_1 := \tilde{\psi}_1, && \hat{\psi}_2 := \tilde{\psi}_2, && \hat{\psi}_3 := L\tilde{\psi}_3, && \hat{\psi}_4 := L\tilde{\psi}_4, \\ &\hat{\psi}_5 := L^{m-1}\tilde{\psi}_5, && \hat{\psi}_6 := \tilde{\psi}_6, && \hat{\psi}_7 := \tilde{\psi}_7. && \end{aligned}$$

Proof. It is obvious that $\hat{A}$ and $\hat{a}$ are measurable functions. Regarding (1.9.8) on V, we have

$$\begin{aligned} p\,\hat{A}(x,u,p) &= pS'(x)\,A(Sx,u,pS'(x)) \geq \nu|pS'(x)|^m - \psi_1(Sx)|u|^m - \psi_2(Sx) \\ &\geq \alpha^m \nu |p|^m - \tilde{\psi}_1(x)|u|^m - \tilde{\psi}_2(x) \end{aligned}$$

The estimates (1.9.9) and (1.9.10) are obvious on V. On U, all three estimates are trivial. Finally, the functions $\hat{\psi}_k$ are in the required spaces by Lemma 2.1.2, part (iv). □

Lemma 2.1.4. *Let $g \in L^q(\partial\Omega)$, $q \in [1,\infty)$, and let $u \in W^{1,m}(\Omega)$ satisfy*

$$\int_\Omega \nabla\eta\, A(x,u,\nabla u) + \int_\Omega \eta\, a(x,u,\nabla u) = \int_{\partial\Omega} \eta\, g$$

for every $\eta \in \mathrm{C}^\infty(\overline{\Omega})$. Then there exist functions h_j in $L^{qN/(N-1)}(H)$ such that

$$\int_H \nabla\eta\, \hat{A}(x,\tilde{u},\nabla\tilde{u}) + \int_H \eta\, \hat{a}(x,\tilde{u},\nabla\tilde{u}) = \int_H \eta\, h_0 + \sum_{j=1}^N \int_H D_j\eta\, h_j$$

for all $\eta \in \mathrm{C}_c^\infty(H)$. Moreover,

$$\|h_j\|_{L^{qN/(N-1)}(H)} \leq c\,\|g\|_{L^q(\partial\Omega)} \tag{2.1.1}$$

for a constant $c \geq 0$ independent of u and g.

Proof. Let $\eta \in \mathrm{C}_c^\infty(H)$ be fixed. By Lemma 2.1.2, on V we have

$$\nabla\tilde{u} = \nabla u(Sx)S'(x) \quad \text{and} \quad \nabla\eta^*(Sx) = \nabla\eta(x)S'(x).$$

Hence, by Lemma 2.1.2 and Theorem 1.2.3, and since $\eta^*|_\Omega \in \mathrm{C}^\infty(\overline{\Omega})$,

$$\begin{aligned} &\int_V \nabla\eta\, \hat{A}(x,\tilde{u},\nabla\tilde{u}) + \int_V \eta\, \hat{a}(x,\tilde{u},\nabla\tilde{u}) \\ &\quad = \int_V \nabla\eta(x)\, S'(x) A\big(Sx, u(Sx), \nabla u(Sx)\big) + \int_V \eta(x)\, a\big(Sx, u(Sx), \nabla u(Sx)\big) \\ &\quad = \int_V \nabla\eta^*(Sx)\, A\big(Sx, u(Sx), \nabla u(Sx)\big) + \int_V \eta^*(Sx)\, a\big(Sx, u(Sx), \nabla u(Sx)\big) \end{aligned}$$

$$= \int_U \nabla\eta^*(x)\, A\big(x, u(x), \nabla u(x)\big) + \int_U \eta^*(x)\, a\big(x, u(x), \nabla u(x)\big)$$
$$= \int_{\partial\Omega} \eta^*\, g = \int_{\partial\Omega\cap H} \eta\, g.$$

Thus

$$\begin{aligned}
&\int_H \nabla\eta\, \hat{A}(x, \tilde{u}, \nabla\tilde{u}) + \int_H \eta\, \hat{a}(x, \tilde{u}, \nabla\tilde{u}) \\
&\quad = \int_U \nabla\eta\, A(x, u, \nabla u) + \int_U \eta\, a(x, u, \nabla u) + \int_V \nabla\eta\, \hat{A}(x, \tilde{u}, \nabla\tilde{u}) + \int_V \eta\, \hat{a}(x, \tilde{u}, \nabla\tilde{u}) \\
&\quad = 2 \int_{\partial\Omega\cap H} \eta\, g.
\end{aligned}$$

Hence the proof is complete once we show that there exist functions h_j in $L^{qN/(N-1)}(H)$ satisfying (2.1.1) and

$$2 \int_{\partial\Omega\cap H} \eta\, g = \int_H \eta\, h_0 + \sum_{j=1}^N \int_H D_j\eta\, h_j \text{ for all } \eta \in \mathrm{C}_c^\infty(H). \tag{2.1.2}$$

Define the linear functional

$$\varphi\colon \mathrm{C}_c^\infty(H) \to \mathbb{R},\ \eta \mapsto 2 \int_{\partial\Omega\cap H} \eta\, g = 2 \int_{\partial U} \eta\, g.$$

Then

$$|\langle\varphi, \eta\rangle| \le 2\, \|g\|_{L^q(\partial\Omega\cap H)}\, \|\eta\|_{L^{q'}(\partial\Omega\cap H)} \le c\, \|g\|_{L^q(\partial\Omega)}\, \|\eta\|_{W^{1,q^\star}(H)}$$

for $q^\star := qN/(qN - N + 1)$ by Hölder's inequality and Theorem 1.3.10, where $c \ge 0$ depends only on Ω and q. Thus φ extends to a bounded linear functional on $W_0^{1,q^\star}(H)$, and

$$\|\varphi\|_{W_0^{1,q^\star}(H)} \le c\, \|g\|_{L^q(\partial\Omega)}\,.$$

Thus by Theorem 1.4.4 there exist $h_j \in L^{(q^\star)'}(\Omega)$ satisfying (2.1.1) and (2.1.2). Since

$$(q^\star)' = \frac{q^\star}{q^\star - 1} = \frac{qN/(qN - N + 1)}{(N-1)/(qN - N + 1)} = \frac{qN}{N-1},$$

this proves the claim. □

Theorem 2.1.5. *Let $\Omega \subset \mathbb{R}^N$ be a bounded Lipschitz domain. Under Assumption 1.9.1, let $q \in [1, \infty]$, and assume $q > \frac{N-1}{m-1}$ if $m \le N$. Then there exist $\alpha \in (0,1)$ and $c \ge 0$ not depending on ψ_2, ψ_4, and ψ_7 such that for $g \in L^q(\partial\Omega)$, every weak solution $u \in W^{1,m}(\Omega)$ of the inhomogeneous Neumann equation*

$$\begin{cases} -\operatorname{div} A(x, u, \nabla u) + a(x, u, \nabla u) = 0 & \text{on } \Omega, \\ \qquad\qquad A(z, u, \nabla u)\cdot\nu = g(z) & \text{on } \partial\Omega \end{cases} \tag{2.1.3}$$

is in $C^{0,\alpha}(\Omega)$ *and*

$$\|u\|_{C^{0,\alpha}(\Omega)} \leq c\Big(\|u\|_{L^m(\Omega)} + \big(\|\psi_4\| + \|\psi_7\| + \|g\|_{L^q(\partial\Omega)}\big)^{1/(m-1)} + \|\psi_2\|^{1/m}\Big). \tag{2.1.4}$$

So in particular u *has a continuous extension to* $\overline{\Omega}$*, which is again Hölder continuous with exponent* α*. Here, the norms of the functions* ψ_2*,* ψ_4*, and* ψ_7 *are to be taken in the spaces indicated in Assumption 1.9.1.*

Proof. In the proof, the symbol c denotes a constant that does not depend on u, g, ψ_2, ψ_4, and ψ_7, but may change its value from line to line.

Let $u \in W^{1,m}(\Omega)$ be a weak solution of (2.1.3). By Theorem 1.9.2, by compactness of $\partial\Omega$, since Hölder continuity is a local property, and since the norm of $C^{0,\alpha}(\Omega)$ does essentially depend only on the local behavior of the function, it suffices to show that for every $z \in \partial\Omega$ there exists a neighborhood H' of z such that $u \in C^{0,\alpha}(H' \cap \Omega)$ with an estimate like (2.1.4).

Let $z \in \partial\Omega$ be fixed and pick H as before. Let H' be a neighborhood of z that is relatively compact in H. Choose $h_j \in L^{qN/(N-1)}(H)$ as in Lemma 2.1.4. Then

$$\int_H \nabla\eta\, \bar{A}(x, \tilde{u}, \nabla\tilde{u}) + \int_H \eta\, \bar{a}(x, \tilde{u}, \nabla\tilde{u}) = 0$$

for every $\eta \in C_c^\infty(H)$ with

$$\bar{A}_j(x,u,p) := \hat{A}_j(x,u,p) - h_j \quad \text{and} \quad \bar{a}(x,u,p) := \hat{a}(x,u,p) - h_0.$$

Using Lemma 2.1.3 and Young's inequality we obtain that for every $\varepsilon > 0$

$$p\,\bar{A}(x,u,p) = p\,\hat{A}(x,u,p) - \sum_{j=1}^{N} p_j h_j \geq \alpha^m \nu |p|^m - \tilde{\psi}_1 |u|^m - \tilde{\psi}_2 - \frac{\varepsilon^m}{m}|p|^m - \frac{1}{m'\varepsilon^{m'}}|h|^{m'},$$

where $h := (h_1, \ldots, h_N)^T$. Let $\varepsilon > 0$ be such that $2\varepsilon^m = m\alpha^m\nu$. Then

$$p\,\bar{A}(x,u,p) \geq \bar{\nu}|p|^m - \tilde{\psi}_1|u|^m - \bar{\psi}_2, \tag{2.1.5}$$

with $\bar{\nu} := \frac{\alpha^m \nu}{2} > 0$ and

$$\left|\bar{\psi}_2\right| \leq \left|\tilde{\psi}_2\right| + c|h|^{m'}.$$

First let $m \leq N$. Then $q > \frac{N-1}{m-1}$ by assumption, hence $\frac{qN}{N-1} > \frac{N}{m-1}$. Thus for sufficiently small $\varepsilon' > 0$,

$$h_j \in L^{\frac{Nm}{(m-\varepsilon')(m-1)}}(H), \ \|h_j\|_{L^{\frac{Nm}{(m-\varepsilon')(m-1)}}(H)} \leq c\,\|g\|_{L^q(\partial\Omega)}\,.$$

We pick $\varepsilon' > 0$ so small that $\frac{N}{m-\varepsilon'} < \frac{N}{m} + \varepsilon$ and $\frac{N}{m-1-\varepsilon'} < \frac{N}{m-1} + \varepsilon$ for ε as in Assumption 1.9.1. Then

$$|h|^{m'} = |h|^{m/(m-1)} \in L^{N/(m-\varepsilon')}(H), \ \left\| |h|^{m'} \right\|_{L^{N/(m-\varepsilon')}(H)} \leq c \, \|g\|_{L^q(\partial\Omega)}^{m/(m-1)}$$

and thus

$$\bar{\psi_2} \in L^{N/(m-\varepsilon')}(H), \ \left\| \bar{\psi_2} \right\|_{L^{N/(m-\varepsilon')}(H)} \leq c\Big(\|\psi_2\|_{L^{N/m+\varepsilon}(\Omega)} + \|g\|_{L^q(\partial\Omega)}^{m/(m-1)} \Big)$$

by Lemma 2.1.2, part (iv). Hence estimate (2.1.5) for $\bar{A}$ is of the type that is required in (1.9.8). Since by (2.1.1)

$$h_j \in L^{N/(m-1-\varepsilon')}(H), \ \|h_j\|_{L^{N/(m-1-\varepsilon')}(H)} \leq c \, \|g\|_{L^q(\partial\Omega)}$$

and in particular

$$h_0 \in L^{N/(m-\varepsilon')}(H), \ \|h_0\|_{L^{N/(m-\varepsilon')}(H)} \leq c \, \|g\|_{L^q(\partial\Omega)} ,$$

estimates (1.9.9) and (1.9.10) are satisfied for $\bar{A}$ and $\bar{a}$, with $\bar{\psi_4} := \tilde{\psi}_4 + \sum_{j=1}^N |h_j|$ and $\bar{\psi_7} := \tilde{\psi}_7 + |h_0|$, and

$$\|\bar{\psi_4}\|_{L^{N/(m-1-\varepsilon')}(H)} \leq c\Big(\|\psi_4\|_{L^{N/(m-1)+\varepsilon}(U)} + \|g\|_{L^q(\partial\Omega)} \Big)$$

and

$$\|\bar{\psi_7}\|_{L^{N/(m-\varepsilon')}(H)} \leq c\Big(\|\psi_7\|_{L^{N/m+\varepsilon}(U)} + \|g\|_{L^q(\partial\Omega)} \Big).$$

Now Theorem 1.9.2 implies that $\tilde{u} \in \mathrm{C}^{0,\alpha}(H')$ for some $\alpha \in (0,1)$ with

$$\begin{aligned} \|\tilde{u}\|_{\mathrm{C}^{0,\alpha}(H')} &\leq c\Big(\|\tilde{u}\|_{L^m(H)} + \big(\|\bar{\psi_4}\| + \|\bar{\psi_7}\| \big)^{1/(m-1)} + \|\bar{\psi_2}\|^{1/m} \Big) \\ &\leq c\Big(\|u\|_{L^m(U)} + \big(\|\psi_4\| + \|\psi_7\| + \|g\|_{L^q(\partial\Omega)} \big)^{1/(m-1)} + \big(\|\psi_2\| + \|g\|_{L^q(\partial\Omega)}^{m/(m-1)} \big)^{1/m} \Big). \end{aligned}$$

Thus $\tilde{u}|_{U'} \in \mathrm{C}^{0,\alpha}(U')$ for $U' := H' \cap \Omega$ with $\|\tilde{u}|_{U'}\|_{\mathrm{C}^{0,\alpha}(U')} \leq \|\tilde{u}\|_{\mathrm{C}^{0,\alpha}(H')}$. Hence the theorem is proved if $m \leq N$, see also Lemma 1.1.3.

Now let $m > N$. Then by (2.1.1)

$$h_j \in L^{N/(N-1)}(H), \ \|h_j\|_{L^{N/(N-1)}(H)} \leq c \, \|g\|_{L^1(\partial\Omega)} .$$

Since $x \mapsto \frac{x}{x-1}$ is monotonically decreasing for $x > 1$, we obtain that

$$h_j \in L^{m/(m-1)}(H), \ \|h_j\|_{L^{m/(m-1)}(H)} \leq c \, \|g\|_{L^q(\partial\Omega)} ,$$

and in particular

$$h_0 \in L^1(H), \ \|h_j\|_{L^1(H)} \leq c \, \|g\|_{L^q(\partial\Omega)} .$$

Moreover,

$$|h|^{m'} = |h|^{m/(m-1)} \in L^1(H), \quad \left\| |h|^{m'} \right\|_{L^1(H)} \le c \|g\|_{L^q(\partial\Omega)}^{m/(m-1)}.$$

Using these estimates, we see as in the case $m \le N$ that $\bar{A}$ and $\bar{a}$ satisfy Assumption 1.9.1, where the functions $\bar{\psi}_2$, $\bar{\psi}_4$, and $\bar{\psi}_7$ can be estimated like before. Again the claim follows from Theorem 1.9.2. □

Corollary 2.1.6. *Let $\Omega \subset \mathbb{R}^N$ be a bounded Lipschitz domain, and let $\varepsilon > 0$. Under Assumption 1.9.1, there exist $\alpha \in (0,1)$ and a constant $c \ge 0$ with the following property. Let f_0 and f_j be functions on Ω, and let g be a function on $\partial\Omega$. If $m \le N$, we assume $f_0 \in L^{N/m+\varepsilon}(\Omega)$, $f_j \in L^{N/(m-1)+\varepsilon}(\Omega)$, and $g \in L^{(N-1)/(m-1)+\varepsilon}(\partial\Omega)$. If $m > N$, we assume $f_0 \in L^1(\Omega)$, $f_j \in L^{m/(m-1)}(\Omega)$, $j = 1, \dots, N$, and $g \in L^1(\partial\Omega)$. Then every weak solution $u \in W^{1,m}(\Omega)$ of (1.9.7) is in $C^{0,\alpha}(\Omega)$ with*

$$\|u\|_{C^{0,\alpha}(\Omega)} \le c\Big(\|u\|_{L^m(\Omega)} + \Big(\|f_0\| + \sum_{j=1}^{N} \|f_j\| + \|g\| \Big)^{1/(m-1)} + 1 \Big), \tag{2.1.6}$$

where the norms of f_0, f_j, and g are to be taken in the aforementioned spaces.

Proof. Define

$$\bar{A}_j(x,u,p) := A_j(x,u,p) - f_j(x) \quad \text{and} \quad \bar{a}(x,u,p) := a(x,u,p) - f_0(x).$$

Then u is a weak solution of

$$\begin{cases} -\operatorname{div} \bar{A}(x,u,\nabla u) + \bar{a}(x,u,\nabla u) = 0 & \text{on } \Omega, \\ \bar{A}(z,u,\nabla u) \cdot \nu = g & \text{on } \partial\Omega. \end{cases}$$

The same arguments as in the proof of Theorem 2.1.5 show that $\bar{A}$ and $\bar{a}$ satisfy Assumption 1.9.1 with

$$\|\bar{\psi}_2\| \le c\Big(\|\psi_2\| + \sum_{j=1}^{N} \|f_j\|^{m/(m-1)} \Big), \qquad \|\bar{\psi}_4\| \le c\Big(\|\psi_4\| + \sum_{j=1}^{N} \|f_j\| \Big),$$
$$\|\bar{\psi}_7\| \le c\Big(\|\psi_7\| + \|f_0\| \Big),$$

where the norms are to be taken in the spaces indicated in Assumption 1.9.1. Thus

$$\Big(\|\bar{\psi}_4\| + \|\bar{\psi}_7\| \Big)^{1/(m-1)} + \|\bar{\psi}_2\|^{1/m} \le c\Big(\Big(\|f_0\| + \sum_{j=1}^{N} \|f_j\| \Big)^{1/(m-1)} + 1 \Big)$$

by Lemma 1.1.3. Now the claim follows from Theorem 2.1.5. □

The proof of Corollary 2.1.6 reveals the following fact.

Remark 2.1.7. In Corollary 2.1.6, assume in addition that $\psi_2 = \psi_4 = \psi_7 = 0$ in Assumption 1.9.1. Then estimate (2.1.6) can be refined to

$$\|u\|_{C^{0,\alpha}(\Omega)} \leq c\Big(\|u\|_{L^m(\Omega)} + \Big(\|f_0\| + \sum_{j=1}^{N} \|f_j\| + \|g\|\Big)^{1/(m-1)}\Big).$$

2.2 Existence of solutions

For the regularity results in Chapter 3, we need that a resolvent operator for the Neumann problem exists, i.e., we need existence of unique weak solutions for the associated Neumann problem. This section is devoted to the study of existence of solutions. As usual in the theory of partial differential equations, we can expect existence of solutions only if we add a sufficiently large lower order term so that the resulting differential operator will be coercive.

2.2.1 Linear equations

We start with linear problems. For those, the Lax-Milgram lemma constitutes a handy tool to prove existence and even uniqueness of weak solutions.

Lemma 2.2.1. *Let $\Omega \subset \mathbb{R}^N$ be a bounded Lipschitz domain. Under Assumption 1.9.3, there exist $\alpha > 0$ and $\omega_0 \geq 0$ such that*

$$\sum_{i,j=1}^{N} \int_\Omega a_{ij} D_i u\, D_j u + \sum_{j=1}^{N} \int_\Omega b_j u\, D_j u + \sum_{i=1}^{N} \int_\Omega c_i D_i u\, u + \int_\Omega d u^2 \geq \alpha \|\nabla u\|^2_{L^2(\Omega)} - \omega_0 \|u\|^2_{L^2(\Omega)}.$$

for all $u \in W^{1,2}(\Omega)$.

Proof. For simplicity we assume that $\sum_j \|b_j + c_j\|^2_{L^q(\Omega)} \neq 0$ and $\|d\|_{L^{q/2}(\Omega)} \neq 0$ since we are going to divide by these expressions. But in fact, if either of these terms equals zero, the respective part of the proof becomes trivial, so this assumption is no real restriction.

Let $u \in W^{1,2}(\Omega)$ be arbitrary. By Assumption 1.9.3,

$$\sum_{i,j=1}^{N} \int_\Omega a_{ij} D_i u\, D_j u \geq \nu \int_\Omega |\nabla u|^2 = \nu \|\nabla u\|^2_{L^2(\Omega)}. \tag{2.2.1}$$

Moreover, by Hölder's and Young's inequalities

$$\begin{aligned}\left|\sum_{j=1}^N \int_\Omega b_j u\, D_j u + \sum_{i=1}^N \int_\Omega c_i D_i u\, u\right| &\le \sum_{j=1}^N \int_\Omega |(b_j + c_j) u\, D_j u| \\ &\le \sum_{j=1}^N \|b_j + c_j\|_{L^q(\Omega)} \|u\|_{L^{2q/(q-2)}(\Omega)} \|D_j u\|_{L^2(\Omega)} \\ &\le N^{1/2}\Big(\sum_{j=1}^N \|b_j + c_j\|^2_{L^q(\Omega)}\Big)^{1/2} \|u\|_{L^{2q/(q-2)}(\Omega)} \|\nabla u\|_{L^2(\Omega)} \\ &\le \frac{\nu}{4}\|\nabla u\|^2_{L^2(\Omega)} + \frac{N}{\nu}\|u\|^2_{L^{2q/(q-2)}(\Omega)} \sum_{j=1}^N \|b_j + c_j\|^2_{L^q(\Omega)} .\end{aligned}$$

Since $W^{1,2}(\Omega)$ is compactly embedded into $L^{2q/(q-2)}(\Omega)$ for $N \ge 2$ by Theorem 1.3.9, for every $\varepsilon > 0$ there exists $c_\varepsilon > 0$ not depending on u such that

$$\|u\|^2_{L^{2q/(q-2)}(\Omega)} \le \varepsilon \|\nabla u\|^2_{L^2(\Omega)} + c_\varepsilon \|u\|^2_{L^2(\Omega)},$$

see Example 1.4.8. For $\varepsilon := \frac{\nu^2}{4N}\Big(\sum_{j=1}^N \|b_j + c_j\|^2_{L^q(\Omega)}\Big)^{-1}$ we obtain that

$$\left|\sum_{j=1}^N \int_\Omega b_j u\, D_j u + \sum_{i=1}^N \int_\Omega c_i D_i u\, u\right| \le \frac{\nu}{2}\|\nabla u\|^2_{L^2(\Omega)} + \frac{N c_\varepsilon}{\nu} \sum_{j=1}^N \|b_j + c_j\|^2_{L^q(\Omega)} \|u\|^2_{L^2(\Omega)}. \quad (2.2.2)$$

In a similar way we see that

$$\begin{aligned}\left|\int_\Omega du^2\right| &\le \|d\|_{L^{q/2}(\Omega)} \|u^2\|_{L^{q/(q-2)}(\Omega)} = \|d\|_{L^{q/2}(\Omega)} \|u\|^2_{L^{2q/(q-2)}(\Omega)} \\ &\le \tilde{\varepsilon} \|d\|_{L^{q/2}(\Omega)} \|\nabla u\|^2_{L^2(\Omega)} + c_{\tilde{\varepsilon}} \|d\|_{L^{q/2}(\Omega)} \|u\|^2_{L^2(\Omega)} \\ &= \frac{\nu}{4}\|\nabla u\|^2_{L^2(\Omega)} + c_{\tilde{\varepsilon}} \|d\|_{L^{q/2}(\Omega)} \|u\|^2_{L^2(\Omega)}\end{aligned} \quad (2.2.3)$$

with $\tilde{\varepsilon} := \frac{\nu}{4}\|d\|^{-1}_{L^{q/2}(\Omega)}$. Combining (2.2.1), (2.2.2), and (2.2.3),

$$\begin{aligned}&\sum_{i,j=1}^N \int_\Omega a_{ij} D_i u\, D_j u + \sum_{j=1}^N \int_\Omega b_j u\, D_j u + \sum_{i=1}^N \int_\Omega c_i D_i u\, u + \int_\Omega du^2 \\ &\quad \ge \frac{\nu}{4}\|\nabla u\|^2_{L^2(\Omega)} - \omega_0 \|u\|^2_{L^2(\Omega)}\end{aligned}$$

for some $\omega_0 \ge 0$ that does not depend on u. □

Proposition 2.2.2. *Let Ω be a bounded Lipschitz domain. Under Assumption 1.9.3, there exists $\omega_0 \ge 0$ such that for all $\omega > \omega_0$ and $r > 1$ we can find $c \ge 0$ with the following property. If $N = 2$, let $f_0 \in L^r(\Omega)$, $f_j \in L^2(\Omega)$, and $g \in L^r(\partial\Omega)$, whereas for $N \ge 3$ let $f_0 \in L^{2N/(N+2)}(\Omega)$, $f_j \in L^2(\Omega)$, and $g \in L^{2(N-1)/N}(\partial\Omega)$. Then there exists a* *unique weak solution* $u \in W^{1,2}(\Omega)$ *of*

$$\begin{cases} -\sum_{j=1}^{N} D_j\Big(\sum_{i=1}^{N} a_{ij} D_i u + b_j u\Big) + \sum_{i=1}^{N} c_i D_i u + du + \omega u = f_0 - \sum_{j=1}^{N} D_j f_j & \text{on } \Omega, \\ \sum_{j=1}^{N}\Big(\sum_{i=1}^{N} a_{ij} D_i u + b_j u\Big)\nu_j = g + \sum_{j=1}^{N} f_j \nu_j & \text{on } \partial\Omega. \end{cases} \tag{2.2.4}$$

Moreover, this solution satisfies

$$\|u\|_{W^{1,2}(\Omega)} \leq c\Big(\|f_0\| + \sum_{j=1}^{N} \|f_j\| + \|g\|\Big), \tag{2.2.5}$$

where the norms are to be taken in the aforementioned spaces.

Proof. For $u, v \in W^{1,2}(\Omega)$, define

$$\mathfrak{a}^\omega(u,v) := \sum_{i,j=1}^{N} \int_\Omega a_{ij} D_i u\, D_j v + \sum_{j=1}^{N} \int_\Omega b_j u\, D_j v + \sum_{i=1}^{N} \int_\Omega c_i D_i u\, v + \int_\Omega (d+\omega) uv. \tag{2.2.6}$$

Then $\mathfrak{a}^\omega$ is a continuous form on $W^{1,2}(\Omega)$, see estimates (2.2.2) and (2.2.3). Take $\omega_0 \geq 0$ as in Lemma 2.2.1. Then

$$\mathfrak{a}^\omega(u,u) \geq \alpha \|\nabla u\|^2_{L^2(\Omega)} + (\omega - \omega_0)\|u\|^2_{L^2(\Omega)} \geq \min\{\alpha, \omega - \omega_0\}\|u\|^2_{W^{1,2}(\Omega)},$$

i.e., $\mathfrak{a}^\omega$ is coercive.

Theorems 1.3.9 and 1.3.10 and Hölder's inequality show that

$$\left|\int_{\partial\Omega} g\, v - \int_\Omega f_0\, v - \sum_{j=1}^{N} \int_\Omega f_j\, D_j v\right| \leq c_1\Big(\|f_0\| + \sum_{j=1}^{N} \|f_j\| + \|g\|\Big)\, \|v\|_{W^{1,2}(\Omega)},$$

where $c_1 \geq 0$ depends only on the function spaces. This means that

$$\varphi(v) := \int_{\partial\Omega} g\, v - \int_\Omega f_0\, v - \sum_{j=1}^{N} \int_\Omega f_j\, D_j v$$

defines a bounded linear functional φ on $W^{1,2}(\Omega)$ with

$$\|\varphi\| \leq c_1\Big(\|f_0\| + \sum_{j=1}^{N} \|f_j\| + \|g\|\Big).$$

Since $\mathfrak{a}^\omega$ is coercive, the Lax-Milgram lemma, i.e., Theorem 1.4.6, implies that there exists a unique $u \in W^{1,2}(\Omega)$ such that

$$\mathfrak{a}^{\omega}(u,v) = \varphi(v) \text{ for all } v \in W^{1,2}(\Omega) \tag{2.2.7}$$

and

$$\|u\|_{W^{1,2}(\Omega)} \leq c_2\|\varphi\| \leq c_1c_2\Big(\|f_0\| + \sum_{j=1}^{N}\|f_j\| + \|g\|\Big)$$

for a constant $c_2 \geq 0$ not depending on u, f_0, f_j, and g. Expanding the definitions of $\mathfrak{a}^{\omega}$ and φ and using that $C^{\infty}(\overline{\Omega})$ is dense in $W^{1,2}(\Omega)$, we see that (2.2.7) is equivalent to u being a weak solution in $W^{1,2}(\Omega)$ of (2.2.4). □

2.2.2 Quasilinear equations

In this section we recall how the theory of monotone or more generally pseudo-monotone operators is used to prove existence of solutions to (1.9.7) provided that apart from growth conditions the leading term also satisfies a monotonicity assumption. For this it will be useful to consider the differential operator as a mapping from $W^{1,m}(\Omega)$ into its dual space, i.e.,

$$\mathcal{A}\colon W^{1,m}(\Omega) \to W^{1,m}(\Omega)', \ \langle \mathcal{A}(u), v\rangle := \int_{\Omega} \nabla v\, A(x,u,\nabla u) + \int_{\Omega} v\, a(x,u,\nabla u), \tag{2.2.8}$$

provided of course that for fixed u this indeed defines a bounded linear functional on $W^{1,m}(\Omega)$.

The following lemma states the rather obvious relation between the operator $\mathcal{A}$ and the differential equation (1.9.7).

Lemma 2.2.3. *Let m be in $(1,\infty)$, let Ω be a bounded Lipschitz domain, and let Assumption 1.9.1 be fulfilled. Let f_0 and f_j be functions on Ω and g be a function on $\partial\Omega$ such that $f_j \in L^{m/(m-1)}(\Omega)$ and*

- *if $m < N$, then $f_0 \in L^{mN/(mN+m-N)}(\Omega)$ and $g \in L^{m(N-1)/(mN-N)}(\partial\Omega)$;*
- *if $m = N$, then $f_0 \in L^{r}(\Omega)$ and $g \in L^{r}(\partial\Omega)$ for some $r > 1$;*
- *if $m > N$, then $f_0 \in L^{1}(\Omega)$ and $g \in L^{1}(\partial\Omega)$.*

Then

$$\varphi\colon W^{1,m}(\Omega) \to \mathbb{R},\ v \mapsto \int_{\Omega} v\, f_0 + \sum_{j=1}^{N}\int_{\Omega} D_j v\, f_j + \int_{\partial\Omega} v\, g$$

defines a bounded linear functional on $W^{1,m}(\Omega)$. Moreover, $u \in W^{1,m}(\Omega)$ is a weak solution of (1.9.7) if and only if $\mathcal{A}(u) = \varphi$.

Proof. The boundedness of φ is an easy consequence of Hölder's inequality and Theorems 1.3.9 and 1.3.10. The equivalence of u being a solution of (1.9.7) to $\mathcal{A}(u) = \varphi$ follows from the definitions combined with Theorem 1.3.6. □

By Lemma 2.2.3, solving (1.9.7) is the same as finding $u \in W^{1,m}(\Omega)$ such that $\mathcal{A}(u) = \varphi$ for an appropriately chosen $\varphi \in W^{1,m}(\Omega)$. The next theorem allows us to apply Theorem 1.5.5 to solve the latter equation.

Assumption 2.2.4. *Let $m \in (1, \infty)$. Let the functions A and a be measurable, and assume that A and a are continuous with respect to u and p. Assume that there exist $\nu, \mu > 0$ and non-negative functions $\psi_1 \in L^{m/(m-1)}(\Omega)$, and $\psi_2 \in L^1(\Omega)$ such that*

$$\begin{aligned} &|A(x,u,p)| \leq \mu|p|^{m-1} + \mu|u|^{m-1} + \psi_1, \qquad |a(x,u,p)| \leq \mu|p|^{m-1} + \mu|u|^{m-1} + \psi_1, \\ &p\, A(x,u,p) \geq \nu|p|^m - \mu|u|^m - \psi_2 \end{aligned}$$

for all $x \in \Omega$, $u \in \mathbb{R}$, and $p \in \mathbb{R}^N$. Moreover, assume that

$$(p - \tilde{p})\Big(A(x,u,p) - A(x,u,\tilde{p})\Big) > 0 \tag{2.2.9}$$

for all $x \in \Omega$, $u \in \mathbb{R}$, and $p \neq \tilde{p} \in \mathbb{R}^N$.

Theorem 2.2.5 (Leray-Lions, [Sho97, Theorem II.6.1]). *Let Ω be a bounded Lipschitz domain. Under Assumption 2.2.4, the operator $\mathcal{A}$ defined as in (2.2.8) is well-defined and pseudo-monotone.*

It will be useful to consider the shifted version

$$\begin{cases} -\operatorname{div} A(x,u,\nabla u) + a(x,u,\nabla u) + \omega|u|^{m-2}u = f_0(x) - \sum\limits_{j=1}^N D_j f_j(x) & \text{on } \Omega \\ A(z,u,\nabla u)\cdot \nu(z) = g(z) + \sum\limits_{j=1}^N f_j(z)\nu_j(z) & \text{on } \partial\Omega, \end{cases} \tag{2.2.10}$$

of equation (1.9.7) for some given $\omega \in \mathbb{R}$.

Theorem 2.2.6. *Let Ω be a bounded Lipschitz domain, and let $m \in (1, \infty)$. Under Assumption 2.2.4, there exists $\omega_0 \geq 0$ such that for every $\omega > \omega_0$ equation (2.2.10) has a weak solution $u \in W^{1,m}(\Omega)$ whenever f_0, f_j, and g are functions as in Lemma 2.2.3. In addition, there exists $c \geq 0$ not depending on u, f_0, f_j, and g such that*

$$\|u\|_{W^{1,m}(\Omega)} \leq c\bigg(\Big(\|f_0\| + \sum_{j=1}^N \|f_j\| + \|g\|\Big)^{1/(m-1)} + 1\bigg). \tag{2.2.11}$$

Here, the norms are to be taken in the spaces indicated in Lemma 2.2.3.

Proof. Let $\mathcal{A}$ be as in (2.2.8), and define for $\omega \geq 0$

$$\langle \mathcal{A}^\omega(u), v\rangle := \langle \mathcal{A}(u), v\rangle + \omega \int_\Omega |u|^{m-2}uv \tag{2.2.12}$$

for $u, v \in W^{1,m}(\Omega)$, which is the operator belonging to the equation (2.2.10) in the sense of Lemma 2.2.3.

By Assumption 2.2.4 and Hölder's inequality

$$\begin{aligned}
&\left|\langle \mathcal{A}^\omega(u), v\rangle\right| \\
&\leq \int_\Omega \Big(\mu|\nabla u|^{m-1} + \mu|u|^{m-1} + \psi_1\Big)|\nabla v| \\
&\qquad + \int_\Omega \Big(\mu|\nabla u|^{m-1} + (\mu+\omega)|u|^{m-1} + \psi_1\Big)|v| \\
&\leq \Big(\mu\|\nabla u\|_{L^m(\Omega)}^{m-1} + (\mu+\omega)\|u\|_{L^m(\Omega)}^{m-1} + \|\psi_1\|_{L^{m/(m-1)}(\Omega)}\Big) \\
&\qquad \cdot \Big(\|\nabla v\|_{L^m(\Omega)} + \|v\|_{L^m(\Omega)}\Big),
\end{aligned}$$

which shows that $\mathcal{A}^\omega$ is bounded.

The operator $\mathcal{A}^\omega$ is pseudo-monotone for every $\omega \in \mathbb{R}$ by Theorem 2.2.5 since its coefficients fulfill Assumption 2.2.4. By Hölder's and Young's inequalities, we have for all $\varepsilon > 0$ and all $u \in W^{1,m}(\Omega)$ that

$$\begin{aligned}
\langle \mathcal{A}(u), u\rangle &= \int_\Omega \nabla u\, A(x,u,\nabla u) + \int_\Omega u\, a(x,u,\nabla u) \\
&\geq \nu \int_\Omega |\nabla u|^m - \mu\int_\Omega |u|^m - \int_\Omega \psi_2 - \mu\int_\Omega |\nabla u|^{m-1}|u| - \mu\int_\Omega |u|^m - \int_\Omega \psi_1|u| \\
&\geq \Big(\nu - \frac{m-1}{m}\varepsilon^{m/(m-1)}\mu\Big)\int_\Omega |\nabla u|^m - \Big(2\mu + \frac{1}{m\varepsilon^m} + \frac{1}{m}\Big)\int_\Omega |u|^m \\
&\qquad - \frac{m-1}{m}\int_\Omega \psi_1^{m/(m-1)} - \int_\Omega \psi_2.
\end{aligned}$$

Thus, if we pick a sufficiently small $\varepsilon > 0$, we find $\nu' > 0$, $\omega_0 \geq 0$, and $d \geq 0$ such that

$$\langle \mathcal{A}(u), u\rangle \geq \nu'\,\|u\|_{W^{1,m}(\Omega)}^m - \omega_0\|u\|_{L^m(\Omega)}^m - d \tag{2.2.13}$$

for all $u \in W^{1,m}(\Omega)$. Thus $\mathcal{A}^\omega$ is coercive for $\omega > \omega_0$.

Let $\omega > \omega_0$. Then we obtain from Theorem 1.5.5 that $\mathcal{A}^\omega$ is surjective. Let φ be the linear functional on $W^{1,m}(\Omega)$ defined in Lemma 2.2.3. Then there exists $u \in W^{1,m}(\Omega)$ such that $\mathcal{A}^\omega(u) = \varphi$, i.e., u is a weak solution of (2.2.10) by Lemma 2.2.3. Moreover,

$$\begin{aligned}
\nu'\,\|u\|_{W^{1,m}(\Omega)}^m - d &\leq \langle \mathcal{A}^\omega(u), u\rangle = \langle \varphi, u\rangle \leq \|\varphi\|_{W^{1,m}(\Omega)'}\,\|u\|_{W^{1,m}(\Omega)} \\
&\leq \frac{m-1}{m\varepsilon^{m/(m-1)}}\,\|\varphi\|_{W^{1,m}(\Omega)'}^{m/(m-1)} + \frac{\varepsilon^m}{m}\,\|u\|_{W^{1,m}(\Omega)}^m\,.
\end{aligned}$$

for every $\varepsilon > 0$. Picking $\varepsilon > 0$ small enough, we find constants $c_1 \geq 0$ and $c_2 \geq 0$ not depending on u or φ such that

$$\|u\|_{W^{1,m}(\Omega)} \leq \Big(c_1 + c_2\|\varphi\|_{W^{1,m}(\Omega)'}^{m/(m-1)}\Big)^{1/m}.$$

By Lemma 1.1.3 and since $\|\varphi\|_{W^{1,m}(\Omega)'}$ can be estimated in terms of the norms of f_0, f_j, and g, this implies (2.2.11). □

In some situations, the monotonicity condition 2.2.9 already guarantees uniqueness of solutions. We cover a particularly important case with the following corollary.

Corollary 2.2.7. *Under the assumptions of Theorem 2.2.6, assume in addition that A does not depend on u itself, i.e., $A(x,u,\nabla u) = A(x,\nabla u)$, and that $a(x,u,\nabla u) = 0$. Then for every $\omega > 0$ there exists a unique solution of* (2.2.10).

Proof. The operator $\mathcal{A}$ is monotone by (2.2.9). In fact,

$$\langle \mathcal{A}(u) - \mathcal{A}(v), u - v\rangle = \int_\Omega \big(\nabla u - \nabla v\big)\big(A(x,\nabla u) - A(x,\nabla v)\big) \geq 0$$

for all u and v in $W^{1,m}(\Omega)$. Hence $\mathcal{A}^\omega$ is strictly monotone for every $\omega > 0$. In fact, it can easily be checked that

$$\int_\Omega \big(|u|^{m-2}u - |v|^{m-2}v\big)(u - v) \geq 0$$

for all u and v in $L^m(\Omega)$ with equality if and only if $u = v$ almost everywhere, compare Lemma 2.2.8. It is now clear that we can pick $\omega_0 = 0$ in the proof of Theorem 2.2.6, i.e., that indeed there exists a solution of (2.2.10) for every $\omega > 0$.

Assume that there exists $\omega > 0$ such that we can find two different weak solutions u and v of (2.2.10). Then $\mathcal{A}^\omega(u) = \mathcal{A}^\omega(v)$ by Lemma 2.2.3. Since $\mathcal{A}^\omega$ is strictly monotone, this implies that

$$0 = \langle \mathcal{A}^\omega(u) - \mathcal{A}^\omega(v), u - v\rangle > 0,$$

a contradiction. □

The following criterion will be useful in order to check that the differential operators in our examples satisfy (2.2.9).

Lemma 2.2.8. *Let w be a function from $[0,\infty)$ to $\mathbb{R}$. Then*

$$(p - \tilde{p})\big(w(|p|)p^T - w(|\tilde{p}|)\tilde{p}^T\big) > 0 \text{ for all row vectors } p \neq \tilde{p} \text{ in } \mathbb{R}^N \tag{2.2.14}$$

if and only if $w(x)x$ is positive and strictly increasing on $(0,\infty)$. Thus in particular (2.2.14) *holds if w is positive and non-decreasing.*

Proof. Assume (2.2.14). For $0 \leq y < x$ we pick $p := xe^T$ and $\tilde{p} := ye^T$, for some unit vector e in $\mathbb{R}^N$. Then

$$(xe^T - ye^T)\big(w(x)xe - w(y)ye\big) = (x - y)\big(w(x)x - w(y)y\big) > 0.$$

Dividing by $x - y$ we see that $w(x)x > w(y)y$, showing that $x \mapsto w(x)x$ is strictly increasing on $[0,\infty)$. For $y = 0$ we obtain in particular that $w(x)x > 0$ if $x > 0$.

Now let $x \mapsto w(x)x$ be strictly increasing and positive on $(0,\infty)$. To prove (2.2.14) we can without loss of generality assume that $w(0) = 0$. For all p and $\tilde{p}$ we have by the Cauchy-Schwarz inequality that

$$\begin{aligned}(p-\tilde{p})\big(w(|p|)\,p^T - w(|\tilde{p}|)\,\tilde{p}^T\big) &= w(|p|)|p|^2 - \big(w(|p|)+w(|\tilde{p}|)\big)\,(p \mid \tilde{p}) + w(|\tilde{p}|)|\tilde{p}|^2 \\ &\geq w(|p|)|p|^2 - \big(w(|p|)+w(|\tilde{p}|)\big)|p|\,|\tilde{p}| + w(|\tilde{p}|)|\tilde{p}|^2 \\ &= \big(|p|-|\tilde{p}|\big)\big(w(|p|)\,|p| - w(|\tilde{p}|)\,|\tilde{p}|\big) \geq 0.\end{aligned}$$

Now assume that the left hand side equals zero. Then we have equality in both estimates. Since $x \mapsto w(x)x$ is strictly increasing, $|p| = |\tilde{p}|$. Moreover, we have either

$$w(|p|) + w(|\tilde{p}|) = 0,$$

i.e., $p = \tilde{p} = 0$, or

$$(p \mid \tilde{p}) = |p|\,|\tilde{p}|,$$

the latter being only possible for $p = \tilde{p}$ since $|p| = |\tilde{p}|$. □

Example 2.2.9. If $N = 2$, let $f_0 \in L^r(\Omega)$, $f_j \in L^2(\Omega)$, and $g \in L^r(\partial\Omega)$ for some $r > 1$. If $N \geq 3$, let $f_0 \in L^{2N/(N+2)}(\Omega)$, $f_j \in L^2(\Omega)$, and $g \in L^{2(N-1)/N}(\partial\Omega)$. Then for every $\omega > 0$ the equation

$$\begin{cases} -\operatorname{div}\Big(\big(1+\arctan(|\nabla u|^2)\big)\nabla u\Big) + \omega u = f_0 - \sum\limits_{j=1}^{N} D_j f_j & \text{on } \Omega, \\ \big(1+\arctan(|\nabla u|^2)\big)\dfrac{\partial u}{\partial \nu} = g + \sum\limits_{j=1}^{N} f_j \nu_j & \text{on } \partial\Omega \end{cases}$$

has a unique weak solution $u \in W^{1,2}(\Omega)$.

Proof. We check Assumption 2.2.4 with $m = 2$ for

$$A(x,u,p) := \big(1+\arctan(|p|^2)\big)p^T \quad \text{and} \quad a(x,u,p) := 0.$$

The only non-obvious property is the monotonicity condition (2.2.9). But for

$$w(x) := 1 + \arctan(x^2)$$

the function w is positive and strictly increasing on $(0,\infty)$. Thus (2.2.9) is a consequence of Lemma 2.2.8. The claim now follows from Corollary 2.2.7. □

Example 2.2.10. Let $m \in (1,\infty)$ and $k \geq 0$. For every $\omega > 0$, the (degenerate or non-degenerate, depending on whether $k = 0$ or not) m-Laplace equation

$$\begin{cases} -\operatorname{div}\Big(\big(k+|\nabla u|^2\big)^{(m-2)/2}\nabla u\Big) + \omega|u|^{m-2}u = f_0 - \sum_{j=1}^{N} D_j f_j & \text{on } \Omega, \\ \big(k+|\nabla u|^2\big)^{(m-2)/2}\dfrac{\partial u}{\partial \nu} = g + \sum_{j=1}^{N} f_j \nu_j & \text{on } \partial\Omega \end{cases}$$

has a unique weak solution $u \in W^{1,m}(\Omega)$ whenever the functions on the right hand side are in the spaces indicated in Lemma 2.2.3.

Proof. We check that Assumption 2.2.4 is fulfilled. Only the monotonicity condition is non-trivial. This, however, follows from Lemma 2.2.8 applied to

$$w(x) := (k + x^2)^{(m-2)/2}.$$

The claim now follows from Corollary 2.2.7. □

2.3 Continuous dependence

For the transition from Neumann boundary conditions to general Robin boundary conditions, i.e., from (1.9.7) to (1.9.6), we will need that the solution depends continuously on the right hand side. More precisely, we even require the resolvent to be locally Hölder continuous as an operator into $W^{1,m}(\Omega)$. The aim of this section is to give sufficient conditions for this assumption to be fulfilled.

A necessary condition for continuous dependence is of course uniqueness of solutions. Thus it is natural to add a term of the form $\omega|u|^{m-2}u$ for some large ω, compare Proposition 2.2.2 and Theorem 2.2.6. We exhibit three classes of differential equations for which we have continuous dependence.

There is not much known about continuous dependence, in particular if $m < 2$. But there are some recent related results in the literature. Lê [Lê07] showed that for $m \in (1, \infty)$ the resolvent of the m-Laplace operator with Dirichlet boundary conditions is locally Hölder continuous from $L^\infty(\Omega)$ to $C^1(\overline{\Omega})$ if the domain is of class $C^{1,\gamma}$ for some $\gamma \in (0, 1)$. This is optimal in the sense that for less regular Ω the solution is not even contained in $C^1(\overline{\Omega})$. In a preprint, Biegert [Bie10] proved that in particular for Lipschitz domains the m-Laplace operator with Dirichlet or Robin boundary conditions is locally Hölder continuous from appropriate L^p-spaces to $L^\infty(\Omega)$ if m is not too close to 1.

2.3.1 Linear equations

For linear equations, the Lipschitz continuous dependence on the right hand side is automatic.

Proposition 2.3.1. *Let $\Omega \subset \mathbb{R}^N$ be a bounded Lipschitz domain. Under Assumption 1.9.3, there exists $\omega_0 \geq 0$ such that for every $\omega > \omega_0$ we can find $c \geq 0$ with the following property. If (f_0, f_j, g) and $(\tilde{f}_0, \tilde{f}_j, \tilde{g})$ are as in Proposition 2.2.2, then (2.2.4) has weak solutions u and $\tilde{u}$ in $W^{1,2}(\Omega)$ for these respective right hand sides, and*

$$\|u - \tilde{u}\|_{W^{1,2}(\Omega)} \leq c\Big(\|f_0 - \tilde{f}_0\| + \sum_{j=1}^{N} \|f_j - \tilde{f}_j\| + \|g - \tilde{g}\|\Big), \tag{2.3.1}$$

where the norms are to be taken in the spaces indicated in Proposition 2.2.2.

Proof. Take ω_0 as in Proposition 2.2.2. Let ω be larger than ω_0. Then, by Proposition (2.2.2), for every right hand side there exists a unique solution in $W^{1,2}(\Omega)$ of (2.2.4). Let u and $\tilde{u}$ be the weak solutions for the right hand sides (f_0, f_j, g) and $(\tilde{f}_0, \tilde{f}_j, \tilde{g})$. Then, by linearity, $u - \tilde{u}$ is the weak solution for $(f_0 - \tilde{f}_0, f_j - \tilde{f}_j, g - \tilde{g})$. Thus (2.3.1) follows from (2.2.5). □

Also the continuous dependence in the Hölder norm is a consequence of the linearity of the equation. In fact, the only issue is to show that Corollary 2.1.6 applies to equation (2.2.4).

Proposition 2.3.2. *Let $\Omega \subset \mathbb{R}^N$ be a bounded Lipschitz domain, and let $q > N$. Under Assumption 1.9.3, there exist $\omega_0 \geq 0$ and $\alpha \in (0,1)$ such that for every $\omega > \omega_0$ we can find $c \geq 0$ with the following property. If $f_0, \tilde{f}_0 \in L^{q/2}(\Omega)$, $f_j, \tilde{f}_j \in L^q(\Omega)$, and $g, \tilde{g} \in L^{q-1}(\partial\Omega)$, then (2.2.4) has weak solutions u and $\tilde{u}$ in $W^{1,2}(\Omega) \cap C^{0,\alpha}(\Omega)$ for the right hand sides (f_0, f_j, g) and $(\tilde{f}_0, \tilde{f}_j, \tilde{g})$, respectively, and*

$$\|u - \tilde{u}\|_{C^{0,\alpha}(\Omega)} \leq c\Big(\|f_0 - \tilde{f}_0\|_{L^{q/2}(\Omega)} + \textstyle\sum_{j=1}^{N} \|f_j - \tilde{f}_j\|_{L^q(\Omega)} + \|g - \tilde{g}\|_{L^{q-1}(\partial\Omega)}\Big). \tag{2.3.2}$$

Proof. Take $\omega_0 \geq 0$ as in Proposition 2.3.1, and let ω be larger than ω_0. Since

$$\frac{q}{2} > \frac{N}{2} \geq \frac{2N}{N+2}, \qquad q > N \geq 2, \qquad q - 1 > N - 1 \geq \frac{2(N-1)}{N},$$

by Proposition 2.3.1 there exists a weak solutions u in $W^{1,2}(\Omega)$ of (2.2.4) for every right hand side (f_0, f_j, g).

We can regard (2.2.4) as a special case of (1.9.7) with

$$A_j(x, u, \nabla u) = \sum_{i=1}^{N} a_{ij} D_i u + b_j u \quad \text{and} \quad a(x, u, \nabla u) = \sum_{i=1}^{N} c_i D_i u + du + \omega u.$$

We check that for these coefficients Assumption 1.9.1 is satisfied for $m = 2$. Since by

Assumption 1.9.3, the Cauchy-Schwarz inequality, and Young's inequality

$$\begin{aligned} p\,A(x,u,p) &= \sum_{i,j=1}^{N} a_{ij}p_ip_j + \sum_{j=1}^{N} b_j u p_j \geq \nu|p|^2 - \Big(\sum_{j=1}^{N} b_j^2\Big)^{1/2} |u|\,|p| \\ &\geq \Big(\nu - \frac{\varepsilon^2}{2}\Big)|p|^2 - \frac{1}{2\varepsilon^2}\sum_{j=1}^{N} b_j^2|u|^2 \end{aligned}$$

for all $\varepsilon > 0$, and since $\sum_{j=1}^N b_j^2 \in L^{q/2}(\Omega)$, with $q > N$, we have (1.9.8) if we pick $\varepsilon > 0$ small enough. Moreover, by the Cauchy-Schwarz inequality,

$$|A_j(x,u,p)| \leq \Big(\sum_{i=1}^{N} \|a_{ij}\|^2_{L^\infty(\Omega)}\Big)^{1/2}|p| + |b_j|\,|u| \leq c_2|p| + |b_j||u|,$$

which shows (1.9.9). Finally,

$$|a(x,u,p)| \leq \Big(\sum_{i=1}^{N} c_i^2\Big)^{1/2} |p| + (|d| + \omega)|u|.$$

Hence (1.9.10) holds since $\big(\sum_{i=1}^N c_i^2\big)^{1/2} \in L^q(\Omega)$ by Lemma 1.1.3 and $|d| + \omega \in L^{q/2}(\Omega)$, where $q > N$.

We have checked Assumption 1.9.1. So by Remark 2.1.7 there exist $\alpha \in (0,1)$ and $c \geq 0$ such that u and $\tilde{u}$ are in $C^{0,\alpha}(\Omega)$ and, since $u - \tilde{u}$ is a weak solution of (2.2.4),

$$\begin{aligned} \|u - \tilde{u}\|_{C^{0,\alpha}(\Omega)} \leq c\Big(&\|u - \tilde{u}\|_{L^2(\Omega)} + \|f_0 - \tilde{f}_0\|_{L^{q/2}(\Omega)} \\ &+ \sum_{j=1}^{N} \|f_j - \tilde{f}_j\|_{L^q(\Omega)} + \|g - \tilde{g}\|_{L^{q-1}(\Omega)}\Big). \end{aligned}$$

Estimating $\|u - \tilde{u}\|_{L^2(\Omega)}$ by (2.3.1), we obtain (2.3.2). □

2.3.2 Regular equations

For quasilinear equations with strictly elliptic, smooth coefficients, continuous dependence on the right hand side can be proved by linearization of the equation. If the coefficients and their derivatives satisfy certain growth bounds or if all solutions are known to have nicely behaved derivatives, we obtain Hölder continuity of the resolvent.

Assumption 2.3.3. *Assume that the functions A and a are measurable and that these functions are differentiable with respect to u and p with bounded derivatives. Assume moreover that there exists $\nu > 0$ such that*

$$\sum_{i,j=1}^{N} D_{p_i}A_j(x,u,p)\xi_i\xi_j \geq \nu|\xi|^2$$

for all $\xi \in \mathbb{R}^N$, $x \in \Omega$, $u \in \mathbb{R}$, *and* $p \in \mathbb{R}^N$.

Theorem 2.3.4. *Let* $\Omega \subset \mathbb{R}^N$ *be a bounded Lipschitz domain,* $N \geq 3$. *Under Assumption 2.3.3 there exists a real number* ω_0 *such that for every* $\omega > \omega_0$ *there exists* $c \geq 0$ *with the following property. Let* f_0 *and* $\tilde{f}_0$ *be in* $L^{2N/(N+2)}(\Omega)$, *let* f_j *and* $\tilde{f}_j$ *be in* $L^2(\Omega)$, *let* g *and* $\tilde{g}$ *be in* $L^{2(N-1)/N}(\partial\Omega)$, *and let* u *and* $\tilde{u}$ *be weak solutions in* $W^{1,2}(\Omega)$ *of*

$$\begin{cases} -\operatorname{div} A(x,u,\nabla u) + a(x,u,\nabla u) + \omega u = f_0 - \sum_{j=1}^{N} D_j f_j & \text{on } \Omega, \\ A(x,u,\nabla u)\cdot\nu = g + \sum_{j=1}^{N} f_j \nu_j & \text{on } \partial\Omega \end{cases} \tag{2.3.3}$$

for these respective right hand sides. Then

$$\|u-\tilde{u}\|_{W^{1,2}(\Omega)} \leq c\Big(\|f_0-\tilde{f}_0\|_{L^{2N/(N+2)}(\Omega)} + \sum_{j=1}^{N}\|f_j-\tilde{f}_j\|_{L^2(\Omega)} + \|g-\tilde{g}\|_{L^{2(N-1)/N}(\partial\Omega)}\Big).$$

Proof. Let $a^\omega(x,u,p) := a(x,u,p) + \omega u$. Set $w_t := u + t(\tilde{u}-u)$ and define

$$I(t;\eta) := \int_\Omega \nabla\eta\, A(x,w_t,\nabla w_t) + \int_\Omega \eta\, a^\omega(x,w_t,\nabla w_t)$$

for $\eta \in C^\infty(\overline{\Omega})$, which exists by Lemma 1.2.7. Then by assumption we have that

$$I(0;\eta) = \int_\Omega \eta\, f_0 + \sum_{j=1}^{N}\int_\Omega D_j\eta\, f_j + \int_{\partial\Omega}\eta\, g$$

and

$$I(1;\eta) = \int_\Omega \eta\, \tilde{f}_0 + \sum_{j=1}^{N}\int_\Omega D_j\eta\, \tilde{f}_j + \int_{\partial\Omega}\eta\, \tilde{g}$$

Hence, by Lemma 1.2.7,

$$\begin{aligned}
&\int_\Omega(\tilde{f}_0-f_0)\eta + \sum_{j=1}^{N}\int_\Omega(\tilde{f}_j-f_j)D_j\eta + \int_{\partial\Omega}(\tilde{g}-g)\eta \\
&= I(1;\eta) - I(0;\eta) = \int_0^1 \frac{\mathrm{d}}{\mathrm{d}t}I(t;\eta)\Big|_{t=s}\,\mathrm{d}s \\
&= \int_\Omega\int_0^1 \nabla\eta\, D_u A(x,w_s,\nabla w_s)\mathrm{d}s\;(\tilde{u}-u) \\
&\qquad + \sum_{i=1}^{N}\int_\Omega\int_0^1 \nabla\eta\, D_{p_i}A(x,w_s,\nabla w_s)\mathrm{d}s\; D_i(\tilde{u}-u) \\
&\qquad + \int_\Omega\int_0^1 \eta\, D_u a^\omega(x,w_s,\nabla w_s)\mathrm{d}s\;(\tilde{u}-u)
\end{aligned}$$

$$+\sum_{i=1}^{N}\int_{\Omega}\int_{0}^{1}\eta\, D_{p_i}a^{\omega}(x,w_s,\nabla w_s)\mathrm{d}s\; D_i(\tilde{u}-u),$$

i.e., $\breve{u} := \tilde{u} - u$ satisfies

$$\sum_{i,j=1}^{N}\int_{\Omega} a_{ij}D_i\breve{u}D_j\eta + \sum_{j=1}^{N}\int_{\Omega} b_j\breve{u}D_j\eta + \sum_{i=1}^{N}\int_{\Omega} c_iD_i\breve{u}\eta + \int_{\Omega} d^{\omega}\breve{u}\eta$$
$$= \int_{\Omega}\breve{f}\eta + \sum_{j=1}^{N}\int_{\Omega}\breve{f}_jD_j\eta + \int_{\partial\Omega}\breve{g}\eta$$

with

$$\breve{f}_0 := \tilde{f}_0 - f_0, \qquad \breve{f}_j := \tilde{f}_j - f_j, \qquad \breve{g} := \tilde{g} - g$$

and

$$a_{ij} := \int_0^1 D_{p_i}A_j(x,w_s,\nabla w_s)\mathrm{d}s, \qquad b_j := \int_0^1 D_uA_j(x,w_s,\nabla w_s)\mathrm{d}s,$$
$$c_i := \int_0^1 D_{p_i}a^{\omega}(x,w_s,\nabla w_s)\mathrm{d}s, \qquad d^{\omega} := \int_0^1 D_ua^{\omega}(x,w_s,\nabla w_s)\mathrm{d}s.$$

Since $\eta \in \mathrm{C}^{\infty}(\overline{\Omega})$ was arbitrary, this shows that $\breve{u}$ solves a linear problem with coefficients $(a_{ij}, b_j, c_i, d^{\omega})$ and right hand side $(\breve{f}_0, \breve{f}_j, \breve{g})$.

We remark that by definition of a^{ω} the functions c_i do in fact not depend on ω and that d^{ω} is of the form $d^{\omega} = d + \omega$ with some function d that does not depend on ω. More precisely

$$c_i = \int_0^1 D_{p_i}a(x,w_s,\nabla w_s)\mathrm{d}s, \qquad d = \int_0^1 D_ua(x,w_s,\nabla w_s)\mathrm{d}s.$$

Thus $\breve{u}$ solves (2.2.4) for the coefficients (a_{ij}, b_j, c_i, d) and the right hand side $(\breve{f}_0, \breve{f}_j, \breve{g})$.

By Assumption 2.3.3, we have

$$\sum_{i,j=1}^{N} D_{p_i}A_j(x,w_s,\nabla w_s)\xi_i\xi_j \geq \nu|\xi|^2$$

for all $\xi \in \mathbb{R}^N$. Integrating this inequality with respect to s we obtain that

$$\sum_{i,j=1}^{N} a_{ij}\xi_i\xi_j \geq \nu|\xi|^2.$$

By a similar argument, the coefficients a_{ij}, b_j, c_i, and d are bounded. Thus Assumption 1.9.3 is satisfied for these coefficients.

Let ω_0 and c be as in Proposition 2.2.2 for the coefficients (a_{ij}, b_j, c_i, d), and let $\omega > \omega_0$.

Then $\check{u} = \tilde{u} - u$ is a weak solution of (2.2.4). Hence Proposition 2.2.2 asserts that

$$\|\check{u}\|_{W^{1,2}(\Omega)} \leq c\Big(\|\check{f}_0\|_{L^{2N/(N+2)}(\Omega)} + \sum_{j=1}^{N} \|\check{f}_j\|_{L^2(\Omega)} + \|\check{g}\|_{L^{2(N-1)/N}(\partial\Omega)}\Big).$$

This is the claim. □

Remark 2.3.5. The same argument proves an analogous statement for $N = 2$ if we require $f_0, \tilde{f}_0 \in L^r(\Omega)$, $f_j, \tilde{f}_j \in L^2(\Omega)$, and $g, \tilde{g} \in L^r(\partial\Omega)$ for some $r > 1$.

Corollary 2.3.6. *If in the situation of Theorem 2.3.4, where $N = 2$ is allowed, we have that Assumption 1.9.1 is fulfilled and that f_0 and $\tilde{f}_0$ are in $L^{q/2}(\Omega)$, f_j and $\tilde{f}_j$ are in $L^q(\Omega)$, and g and $\tilde{g}$ are in $L^{q-1}(\partial\Omega)$ for some $q > N$, then u and $\tilde{u}$ are in $C^{0,\alpha}(\Omega)$ for some $\alpha \in (0,1)$ and*

$$\|u - \tilde{u}\|_{C^{0,\alpha}(\Omega)} \leq c\Big(\|f_0 - \tilde{f}_0\|_{L^{q/2}(\Omega)} + \sum_{j=1}^{N} \|f_j - \tilde{f}_j\|_{L^q(\Omega)} + \|g - \tilde{g}\|_{L^{q-1}(\partial\Omega)}\Big),$$

where the constant c may be different from the one in Theorem 2.3.4.

Proof. The Hölder continuity of the solutions follows from Theorem 2.1.5. As in the proof of Theorem 2.3.4 we see that $\check{u} := u - \tilde{u}$ solves a strictly elliptic linear problem. Now we proceed by using Proposition 2.3.2 instead of Proposition 2.2.2. Comparing $\check{u}$ with the zero function, the claim follows. □

Example 2.3.7. Consider the quasilinear Neumann problem

$$\begin{cases} -\operatorname{div}\Big(\big(1 + \arctan(|\nabla u|^2)\big)\nabla u\Big) + \omega u = f_0 - \sum_{j=1}^{N} D_j f_j & \text{on } \Omega, \\ \big(1 + \arctan(|\nabla u|^2)\big)\dfrac{\partial u}{\partial \nu} = g + \sum_{j=1}^{N} f_j \nu_j & \text{on } \partial\Omega \end{cases}$$

on a bounded Lipschitz domain $\Omega \subset \mathbb{R}^N$, $N \geq 3$, for fixed $\omega > 0$. Then there exist $\alpha \in (0,1)$ and a constant $c > 0$ such that if u and $\tilde{u}$ are the unique solutions for right hand sides (f_0, f_j, g) and $(\tilde{f}_0, \tilde{f}_j, \tilde{g})$, which exist by Example 2.2.9, then

$$\|u - \tilde{u}\|_{W^{1,2}(\Omega)} \leq c\Big(\|f_0 - \tilde{f}_0\|_{L^{2N/(N+2)}(\Omega)} + \sum_{j=1}^{N} \|f_j - \tilde{f}_j\|_{L^2(\Omega)} + \|g - \tilde{g}\|_{L^{2(N-1)/N}(\partial\Omega)}\Big).$$

and

$$\|u - \tilde{u}\|_{C^{0,\alpha}} \leq c\big(\|f_0 - \tilde{f}_0\|_{L^{q/2}(\Omega)} + \sum_{j=1}^{N} \|f_j - \tilde{f}_j\|_{L^q(\Omega)} + \|g - \tilde{g}\|_{L^{q-1}(\partial\Omega)}\big),$$

provided the right hand sides are in the corresponding spaces.

Proof. We are in the situation of Theorem 2.3.4 for

$$A(x,u,p) = \Big(1+\arctan(|p|^2)\Big)p^T \quad \text{and} \quad a(x,u,p) = 0.$$

We check Assumption 2.3.3. Obviously, there is no problem regarding differentiability of the coefficients and the boundedness of the derivatives. So it only remains to analyze the ellipticity condition for

$$D_{p_i}A_j = \frac{2p_ip_j}{1+|p|^4} + \mathbb{1}_{\{i=j\}}\Big(1+\arctan(|p|^2)\Big).$$

For this we notice that

$$\sum_{i,j=1}^{N} D_{p_i}A_j(x,u,p)\xi_i\xi_j = \frac{2}{1+|p|^4}\,(\xi \mid p)^2 + \Big(1+\arctan(|p|^2)\Big)|\xi|^2 \tag{2.3.4}$$

can be estimated from below by $|\xi|^2$. We have checked Assumption 2.3.3. Assumption 1.9.1 is trivial for this example. Hence the claims follow from Theorem 2.3.4 and Corollary 2.3.6; we merely have to notice that in this situation we may pick $\omega_0 = 0$. □

Assumption 2.3.3 is very restrictive, for example allowing only linear growth of A and a with respect to u and p. Note, however, that the only arguments for which these estimates are used in the proof are convex combinations of the solutions. Hence if we know a priori that the solutions and their gradients are bounded, we can relax the assumption.

Assumption 2.3.8. *Assume that the functions A and a are measurable and that these functions are differentiable with respect to u and p. Assume that whenever $K_0 \subset \mathbb{R}$ and $K_1 \subset \mathbb{R}^N$ are compact sets, then the derivatives of A and a are bounded on $\Omega \times K_0 \times K_1$, and there exists $\nu > 0$ such that*

$$\sum_{i,j=1}^{N} D_{p_i}A_j(x,u,p)\xi_i\xi_j \geq \nu|\xi|^2$$

for all $\xi \in \mathbb{R}^N$, $x \in \Omega$, $u \in K_0$, and $p \in K_1$.

Corollary 2.3.9. *Let $\Omega \subset \mathbb{R}^N$ be a bounded Lipschitz domain, $N \geq 3$. Under Assumption 2.3.8 there exists a real number ω_0 such that for every $\omega > \omega_0$ and $M \geq 0$ there exists $c \geq 0$ with the following property. Let f_0 and $\tilde{f}_0$ be in $L^{2N/(N+2)}(\Omega)$, let f_j and $\tilde{f}_j$ be in $L^2(\Omega)$, and let g and $\tilde{g}$ be in $L^{2(N-1)/N}(\partial\Omega)$. Assume that u and $\tilde{u}$ are weak solutions in $W^{1,2}(\Omega)$ of (2.3.3) for these right hand sides, respectively, such that $|u|$, $|\tilde{u}|$, $|\nabla u|$, and $|\nabla\tilde{u}|$ are bounded by M. Then*

$$\|u-\tilde{u}\|_{W^{1,2}(\Omega)} \leq c\Big(\|f_0-\tilde{f}_0\|_{L^{2N/(N+2)}(\Omega)} + \sum_{j=1}^{N}\|f_j-\tilde{f}_j\|_{L^2(\Omega)} + \|g-\tilde{g}\|_{L^{2(N-1)/N}(\partial\Omega)}\Big).$$

Remark 2.3.10. Remark 2.3.5 and Corollary 2.3.6 carry over to the situation of Corollary 2.3.9.

In order to apply Corollary 2.3.9, we need a priori estimates for solutions and their gradients. Under some rather mild conditions on the coefficients, gradient estimates can be obtained from a priori bounds for the solution itself.

Assumption 2.3.11. *Let $m \in (1, \infty)$. Assume that a is measurable, that A is continuously differentiable with respect to all of its arguments, and that $A_j = D_{p_j} F$ for some differentiable function $F = F(x, u, p)$. Assume moreover that for every $M_0 > 0$ there exist positive constants μ and ν such that*

$$\begin{aligned} pA(x,u,p) &\geq \nu |p|^m - \mu, & & \\ |A(x,u,p)| &\leq \mu(1+|p|)^{m-1}, & |D_{x_j} A(x,u,p)| &\leq \mu(1+|p|)^m, \\ |D_u A(x,u,p)| &\leq \mu(1+|p|)^{m-1}, & |a(x,u,p)| &\leq \mu(1+|p|)^m, \end{aligned}$$

and

$$\nu(1+|p|)^{m-2}|\xi|^2 \leq \sum_{i,j=1}^{N} D_{p_i} A_j(x,u,p)\xi_i\xi_j \leq \mu(1+|p|)^{m-2}|\xi|^2 \tag{2.3.5}$$

hold for all $x \in \Omega$, $|u| \leq M_0$, $p \in \mathbb{R}^N$, and $\xi \in \mathbb{R}^N$.

Lemma 2.3.12 ([Lie83, Theorem 4.5 and Example 4 in §6])**.** *Let Ω be of class C^2. Under Assumptions 2.3.11, let $f \in L^\infty(\Omega)$ and $g \in L^\infty(\partial\Omega)$. Let $u \in W^{1,m}(\Omega) \cap L^\infty(\Omega)$ be a weak solution of*

$$\begin{cases} -\operatorname{div} A(x,u,\nabla u) + a(x,u,\nabla u) = f \text{ in } \Omega, \\ \qquad\qquad\qquad\quad A(x,u,\nabla u)\cdot\nu = g \text{ on } \partial\Omega. \end{cases} \tag{2.3.6}$$

Then u is in $W^{1,\infty}(\Omega)$, and there exists an upper bound for ∇u that depends only on the domain, the coefficients, and (pointwise) upper bounds for u and the right hand side.

For the following theorem, we first apply Lemma 2.3.12 to obtain a bound for the gradient in terms of an upper estimate for the functions on the right hand side. Under Assumption 2.3.8, this suffices to obtain continuous dependence. However, since the constant for this dependence depends on the bound for the gradient, the Hölder continuity is only local and from the space of bounded right hand sides. Hence there is no loss in formulating the dependence estimate directly for the Hölder norm, since we have no result for unbounded right hand sides anyway.

Theorem 2.3.13. *Let Ω be of class C^2, and let $m \geq 2$. Under Assumptions 1.9.1, 2.3.8, and 2.3.11, there exists ω_0 such that for all $\omega > \omega_0$, $M \geq 0$, and $q > N$ there exists $\alpha \in (0,1)$ such that if f and $\tilde{f}$ are in $L^\infty(\Omega)$, g and $\tilde{g}$ are in $L^\infty(\partial\Omega)$, and u and $\tilde{u}$ are weak solutions in $W^{1,m}(\Omega)$ of*

$$\begin{cases} -\operatorname{div} A(x,u,\nabla u) + a(x,u,\nabla u) + \omega u = f & \text{on } \Omega, \\ \qquad\qquad\qquad\qquad\quad A(x,u,\nabla u)\cdot\nu = g & \text{on } \partial\Omega \end{cases}$$

for the respective right hand sides satisfying $\|u\|_{L^m(\Omega)} \le M$ *and* $\|\tilde{u}\|_{L^m(\Omega)} \le M$*, then*

$$\|u - \tilde{u}\|_{C^{0,\alpha}(\Omega)} \le c\Big(\|f - \tilde{f}\|_{L^{q/2}(\Omega)} + \|g - \tilde{g}\|_{L^{q-1}(\partial\Omega)}\Big)$$

with a constant $c \ge 0$ *that depends only on the coefficients,* Ω*,* m*,* ω*,* M*,* q*, and (pointwise) upper bounds for* f*,* $\tilde{f}$*,* g*, and* $\tilde{g}$*.*

Proof. By Corollary 2.1.6, the functions u and $\tilde{u}$ and thus, by Lemma 2.3.12, also their gradients are bounded. More precisely, the upper bound depends on ω only via a lower bound, which we may take as zero, as can be seen from the arguments in [Lie92, Theorem 3.6]. Now we can take ω_0 as in Corollary 2.3.9, or, more precisely, as in Remark 2.3.10, corresponding to these bounds for u, $\tilde{u}$, ∇u, and $\nabla\tilde{u}$, so that the variant of Corollary 2.3.6 corresponding to Corollary 2.3.9 yields the result. □

Remark 2.3.14. Theorem 2.3.13 remains true also for $1 < m < 2$. We had to restrict to $m \ge 2$ to make sure that the non-divergence part $a(x, u, \nabla u) + \omega u$ satisfies the assumptions of Corollary 2.1.6 and Lemma 2.3.12. But as already mentioned in the proof of Theorem 2.3.13, the gradient bound does in fact depend only on a lower bound for the non-divergence part, and the same is true for the bound of the functions itself, compare [LU68, §4.7]. So also for $m < 2$ we can find a uniform estimate for all $\omega \ge 0$, and this was the only relevant step in the proof.

Remark 2.3.15. In the applications, we will easily obtain a priori bounds for $\|u\|_{L^m(\Omega)}$ from Theorem 2.2.6. However, in a rather general situation, e.g., if $a(x, u, \nabla u) = 0$ and $\omega > 0$, such a bound is automatic. For a detailed analysis, see [Lie83, §3].

Example 2.3.16. Let $m \ge 2$, $k > 0$, and $\omega > 0$. Consider the non-degenerate m-Laplace equation

$$\begin{cases} -\operatorname{div}\Big(\big(1 + |\nabla u|^2\big)^{(m-2)/2}\nabla u\Big) + k|u|^{m-2}u + \omega u = f \text{ in } \Omega, \\ \qquad\qquad\qquad \big(1 + |\nabla u|^2\big)^{(m-2)/2}\dfrac{\partial u}{\partial \nu} = g \text{ on } \partial\Omega \end{cases} \tag{2.3.7}$$

on a domain Ω of class C^2. Let u and $\tilde{u}$ be the unique weak solutions in $W^{1,m}(\Omega)$ of this equation for bounded right hand sides f, g and $\tilde{f}$, $\tilde{g}$, respectively, which exist by Example 2.2.10. Then for every $q > N$ there exist $\alpha \in (0, 1)$ and a constant c that depend only on Ω, m, k, ω, q, and upper bounds for the right hand sides such that

$$\|u - \tilde{u}\|_{C^{0,\alpha}(\Omega)} \le c\Big(\|f - \tilde{f}\|_{L^{q/2}(\Omega)} + \|g - \tilde{g}\|_{L^{q-1}(\partial\Omega)}\Big).$$

Proof. We check the assumptions of Theorem 2.3.13 for

$$A(x, u, p) := \big(1 + |p|^2\big)^{(m-2)/2}p^T \quad \text{and} \quad a(x, u, p) = k|u|^{m-2}u.$$

Using Theorem 2.2.6, we can estimate $\|u\|_{L^m(\Omega)}$ in terms of the bounds for the right hand side, and an inspection of the proof shows that this estimate is uniform in $\omega \geq 0$. Since

$$\begin{aligned}
p\,A(x,u,p) &= \left(1+|p|^2\right)^{(m-2)/2}|p|^2 \geq |p|^m,\\
|A(x,u,p)| &\leq 2^{(m-2)/2}|p| + 2^{(m-2)/2}|p|^{m-1} \leq c(|p|^{m-1}+1),\\
|a(x,u,p)| &\leq |k|\,|u|^{m-1},
\end{aligned}$$

for some $c \geq 0$, Assumption 1.9.1 is satisfied.

Regarding Assumption 2.3.11, it is an easy calculation to verify that $A_j = D_{p_j}F$ for

$$F(x,u,p) := \frac{1}{m}(1+|p|^2)^{m/2}.$$

Moreover,

$$D_{p_i}A_j(x,u,p) = (m-2)\left(1+|p|^2\right)^{(m-4)/2} p_i p_j + \mathbb{1}_{\{i=j\}}\left(1+|p|^2\right)^{(m-2)/2},$$

so that

$$\sum_{i,j=1}^{N} D_{p_i}A_j(x,u,p)\xi_i\xi_j = (m-2)\left(1+|p|^2\right)^{(m-4)/2}(\xi \mid p)^2 + \left(1+|p|^2\right)^{(m-2)/2}|\xi|^2.$$

Using Lemma 1.1.3, this implies (2.3.5). Since A does not depend on x or u, the remaining estimates in Assumption 2.3.11 are trivial.

Finally, Assumption 2.3.8 is clearly fulfilled. Hence the claim follows from Theorem 2.3.13; we merely have to note that in this situation we may pick $\omega_0 = 0$. □

2.3.3 Uniformly monotone equations

The results about existence of solutions in Section 2.2 were based on the rather mild monotonicity condition (2.2.9). As seen in Corollary 2.2.7, this condition already guarantees the uniqueness of solutions if the coefficients have some simple structure, e.g., if $A(x,u,\nabla u) = A(x,\nabla u)$ and $a(x,u,\nabla u) = |u|^{m-2}u$.

If in addition we assume that a stronger structure condition than (2.2.9) is fulfilled, we have even continuous dependence on the right hand side. This approach requires no additional smoothness or non-degeneracy assumptions on the coefficients.

Assumption 2.3.17. *Let $m \in [2,\infty)$. Let the functions A and a be measurable, and assume that they are continuous with respect to u and p. Let A and a be of the form $A(x,u,p) = A(x,p)$ and $a(x,u,p) = a(x,u)$. Assume that there exist $\nu, \mu > 0$ and non-negative functions $\psi_1 \in L^{m/(m-1)}(\Omega)$ and $\psi_2 \in L^1(\Omega)$ such that*

$$|A(x,p)| \leq \mu|p|^{m-1} + \psi_1, \quad |a(x,u)| \leq \mu|u|^{m-1} + \psi_1, \quad \textit{and} \quad p\,A(x,p) \geq \nu|p|^m - \psi_2$$

for all $x \in \Omega$, $u \in \mathbb{R}$, and $p \in \mathbb{R}^N$. Assume also that

$$(p-\tilde{p})\Big(A(x,p)-A(x,\tilde{p})\Big) \geq \nu|p-\tilde{p}|^m \tag{2.3.8}$$

and

$$(u-\tilde{u})\Big(a(x,u)-a(x,\tilde{u})\Big) \geq \nu|u-\tilde{u}|^m \tag{2.3.9}$$

for all $x \in \Omega$, p and $\tilde{p}$ in $\mathbb{R}^N$, and u and $\tilde{u}$ in $\mathbb{R}$.

Theorem 2.3.18. *Let Ω be a bounded Lipschitz domain. Under Assumption 2.3.17, equation (1.9.7) has a weak solution $u \in W^{1,m}(\Omega)$ for every right hand side in the spaces indicated in Lemma 2.2.3. Moreover, there exists a constant $c \geq 0$ with the following property. If u and $\tilde{u}$ are solutions of (1.9.7) for right hand sides (f_0, f_j, g) and $(\tilde{f}_0, \tilde{f}_j, \tilde{g})$, then*

$$\|u-\tilde{u}\|_{W^{1,m}(\Omega)} \leq c\Big(\|f_0-\tilde{f}_0\| + \sum_{j=1}^N \|f_j-\tilde{f}_j\| + \|g-\tilde{g}\|\Big)^{1/(m-1)}, \tag{2.3.10}$$

where the norms are to be taken in the spaces indicated in Lemma 2.2.3. In particular, the solution for a given right hand side is unique.

Proof. Let $\mathcal{A}$ be the operator defined in (2.2.8). We use Theorem 1.5.5 to conclude that $\mathcal{A}$ is surjective. Apart from the property

$$\lim_{\|u\|_{W^{1,m}(\Omega)}\to\infty} \frac{\langle \mathcal{A}(u), u\rangle}{\|u\|_{W^{1,m}(\Omega)}} = \infty, \tag{2.3.11}$$

the assumptions of Theorem 1.5.5 can be checked for $\mathcal{A}$ like in the proof of Theorem 2.2.6 since Assumption 2.2.4 is satisfied. Thus it only remains to prove (2.3.11). By assumption and Young's inequality,

$$\begin{aligned}\nabla u\, A(x,\nabla u) &= (\nabla u - 0)(A(x,\nabla u) - A(x,0)) + \nabla u\, A(x,0) \geq \nu\,|\nabla u|^m - \psi_1\,|\nabla u| \\ &\geq \left(\nu - \frac{\varepsilon^m}{m}\right)|\nabla u|^m - \frac{m-1}{m\varepsilon^{m/(m-1)}}\psi_1^{m/(m-1)}\end{aligned}$$

and similarly

$$u\,a(x,u) \geq \left(\nu - \frac{\varepsilon^m}{m}\right)|u|^m - \frac{m-1}{m\varepsilon^{m/(m-1)}}\psi_1^{m/(m-1)}$$

for every $\varepsilon > 0$, i.e.,

$$\langle \mathcal{A}(u), u\rangle \geq \alpha\int_\Omega |\nabla u|^m + \alpha\int_\Omega |u|^m - \int_\Omega c\psi_1^{m/(m-1)} = \alpha\|u\|^m_{W^{1,m}(\Omega)} - c\|\psi_1\|^{m/(m-1)}_{L^{m/(m-1)}(\Omega)} \tag{2.3.12}$$

for some $\alpha > 0$ and some constant $c \geq 0$, if we pick $\varepsilon > 0$ small enough. Since $m > 1$, this proves (2.3.11) and hence the existence of solutions.

Now let u and $\tilde{u}$ be solutions for right hand sides (f_0, f_j, g) and $(\tilde{f}_0, \tilde{f}_j, \tilde{g})$, and let φ

and $\tilde{\varphi}$ denote the corresponding functionals as in Lemma 2.2.3. Then $\mathcal{A}(u) = \varphi$ and $\mathcal{A}(\tilde{u}) = \tilde{\varphi}$. Thus

$$\begin{aligned}\|\varphi - \tilde{\varphi}\|_{W^{1,m}(\Omega)'} \|u - \tilde{u}\|_{W^{1,m}(\Omega)} &\geq \langle \varphi - \tilde{\varphi}, u - \tilde{u}\rangle = \langle \mathcal{A}(u) - \mathcal{A}(\tilde{u}), u - \tilde{u}\rangle \\ &= \int_\Omega (\nabla u - \nabla \tilde{u})\big(A(x, \nabla u) - A(x, \nabla \tilde{u})\big) + \int_\Omega (u - \tilde{u})\big(a(x, u) - a(x, \tilde{u})\big) \\ &\geq \nu \int_\Omega |\nabla u - \nabla \tilde{u}|^m + \nu \int_\Omega |u - \tilde{u}|^m = \nu \|u - \tilde{u}\|^m_{W^{1,m}(\Omega)}.\end{aligned}$$

Thus

$$\|u - \tilde{u}\|_{W^{1,m}(\Omega)} \leq \nu^{-1/(m-1)} \|\varphi - \tilde{\varphi}\|^{1/(m-1)}.$$

Since $\varphi - \tilde{\varphi}$ corresponds to the difference of the right hand sides and the embedding of L^p-spaces into $W^{1,m}(\Omega)'$ which is described by Lemma 2.2.3 is continuous, this proves (2.3.10). □

Theorem 2.3.18 says that the solution depends on the right hand side in a Hölder continuous way in the norm of $W^{1,m}(\Omega)$. Since we already have Corollary 2.1.6 at hand, it is easy to deduce from Theorem 2.3.18 that the dependence on the right hand side is locally Hölder continuous even with respect to the norm of $C^{0,\alpha}(\Omega)$ for some $\alpha \in (0,1)$. However, we have no good control on the Hölder exponents.

Corollary 2.3.19. *Let Ω be a bounded Lipschitz domain. Under Assumptions 2.3.17 and 1.9.1, there exist α and θ in $(0,1)$ and a constant $c \geq 0$ with the following property. Let (f_0, f_j, g) and $(\tilde{f}_0, \tilde{f}_j, \tilde{g})$ be in the spaces indicated in Corollary 2.1.6. Then the unique weak solutions u and $\tilde{u}$ in $W^{1,m}(\Omega)$ of (1.9.7) for these right hand sides are in $C^{0,\alpha}(\Omega)$ and satisfy*

$$\begin{aligned}\|u - \tilde{u}\|_{C^{0,\alpha}} \leq c\Big(\|f_0\| + \|\tilde{f}_0\| + \textstyle\sum_{j=1}^N \|f_j\| + \sum_{j=1}^N \|\tilde{f}_j\| + \|g\| + \|\tilde{g}\| + 1\Big)^{\theta/(m-1)} \\ \cdot \Big(\|f_0 - \tilde{f}_0\| + \textstyle\sum_{j=1}^N \|f_j - \tilde{f}_j\| + \|g - \tilde{g}\|\Big)^{(1-\theta)/(m-1)}\end{aligned}$$

with the norms taken in the mentioned spaces.

Proof. Existence and uniqueness of u and $\tilde{u}$ follow from Theorem 2.3.18. Since Assumption 2.3.17 implies Assumption 2.2.4,

$$\begin{aligned}&\|u\|_{L^m(\Omega)} + \|\tilde{u}\|_{L^m(\Omega)} \\ &\quad\leq c\Big(\|f_0\| + \|\tilde{f}_0\| + \sum_{j=1}^N \|f_j\| + \sum_{j=1}^N \|\tilde{f}_j\| + \|g\| + \|\tilde{g}\| + 1\Big)^{1/(m-1)}\end{aligned} \tag{2.3.13}$$

for some constant $c \geq 0$ by (2.2.11) and Lemma 1.1.3; note that we can pick $\omega = 0$ in Lemma 1.1.3 thanks to Assumption 2.3.17, compare estimate (2.3.12). By Corollary 2.1.6, the solutions are in $C^{0,\alpha'}(\Omega)$ for some $\alpha' \in (0,1)$, and, by combining (2.1.6) with (2.3.13),

$$\|u\|_{C^{0,\alpha'}(\Omega)} + \|\tilde{u}\|_{C^{0,\alpha'}(\Omega)} \leq c\Big(\|f_0\| + \|\tilde{f}_0\| + \sum_{j=1}^N \|f_j\| + \sum_{j=1}^N \|\tilde{f}_j\| + \|g\| + \|\tilde{g}\| + 1\Big)^{1/(m-1)} \tag{2.3.14}$$

for some constant $c \geq 0$. Let $\alpha \in (0,\alpha')$ be arbitrary. By Corollary 1.6.2 there exist $\theta \in (0,1)$ and $c \geq 0$ such that

$$\|u - \tilde{u}\|_{C^{0,\alpha}} \leq c\|u - \tilde{u}\|^{\theta}_{C^{0,\alpha'}} \|u - \tilde{u}\|^{1-\theta}_{L^m(\Omega)}. \tag{2.3.15}$$

The claim follows by plugging (2.3.10) and (2.3.14) into (2.3.15). □

Example 2.3.20. If $m \in [2,\infty)$ and $\omega > 0$, then the equation

$$\begin{cases} -\operatorname{div}\big(|\nabla u|^{m-2}\nabla u\big) + \omega|u|^{m-2}u = f_0 - \sum_{j=1}^N D_j f_j, & \text{on } \Omega, \\ |\nabla u|^{m-2}\dfrac{\partial u}{\partial \nu} = g + \sum_{j=1}^N f_j \nu_j, & \text{on } \partial\Omega \end{cases}$$

fits into the framework of Theorem 2.3.18 and Corollary 2.3.19.

Proof. The growth bounds of Assumption 2.3.17 are trivial for this example, and the monotonicity condition is the statement of Lemma 1.1.4. □

The proof of Theorem 2.3.18 goes through also for $m \in (1,2)$. In that case, however, by (2.3.10), the resolvent operator of (1.9.7) mapping a given right hand side to the corresponding weak solution in $W^{1,m}(\Omega)$ is Fréchet differentiable at every point with derivative zero, hence constant, which is of course absurd. Thus an analogue of Assumption 2.3.17 can never be satisfied for some $m < 2$.

However, if we relax conditions (2.3.8) and (2.3.9), we can still obtain results for $m \leq 2$ that are similar to Theorem 2.3.18.

Assumption 2.3.21. *Let $m \in (1,2]$. Let the functions A and a be measurable, and assume that they are continuous with respect to u and p. Let A and a be of the form $A(x,u,p) = A(x,p)$ and $a(x,u,p) = a(x,u)$. Assume that there exist $\nu, \mu > 0$ and functions $\psi_1 \in L^{m/(m-1)}(\Omega)$ and $\psi_2 \in L^1(\Omega)$ such that*

$$|A(x,p)| \leq \mu|p|^{m-1} + \psi_1, \quad |a(x,u)| \leq \mu|u|^{m-1} + \psi_1, \quad \text{and} \quad p\,A(x,p) \geq \nu|p|^m - \psi_2$$

for all $x \in \Omega$, $u \in \mathbb{R}$, and $p \in \mathbb{R}^N$. Assume also that

$$(p - \tilde{p})\big(A(x,p) - A(x,\tilde{p})\big) \geq \nu(|p| + |\tilde{p}|)^{m-2}|p - \tilde{p}|^2 \tag{2.3.16}$$

and

$$(u-\tilde{u})\big(a(x,u)-a(x,\tilde{u})\big) \geq \nu(|u|+|\tilde{u}|)^{m-2}|u-\tilde{u}|^2 \tag{2.3.17}$$

for all $x \in \Omega$, p and $\tilde{p}$ in $\mathbb{R}^N$, and u and $\tilde{u}$ in $\mathbb{R}$.

Theorem 2.3.22. *Let Ω be a bounded Lipschitz domain. Under Assumption 2.3.21, equation (1.9.7) has a weak solution $u \in W^{1,m}(\Omega)$ for every right hand side in the spaces indicated in Lemma 2.2.3. Moreover, there exists a constant $c \geq 0$ with the following property. If u and $\tilde{u}$ are solutions of (1.9.7) for right hand sides (f_0, f_j, g) and $(\tilde{f}_0, \tilde{f}_j, \tilde{g})$, then*

$$\begin{aligned}\|u-\tilde{u}\|_{W^{1,m}(\Omega)} \leq c\Big(\|f_0\| + \|\tilde{f}_0\| + \textstyle\sum_{j=1}^N \|f_j\| + \sum_{j=1}^N \|\tilde{f}_j\| + \|g\| + \|\tilde{g}\| + 1\Big)^{\frac{2-m}{m-1}} \\ \cdot \Big(\|f_0 - \tilde{f}_0\| + \textstyle\sum_{j=1}^N \|f_j - \tilde{f}_j\| + \|g-\tilde{g}\|\Big),\end{aligned} \tag{2.3.18}$$

where the norms are to be taken in the spaces indicated in Lemma 2.2.3. In particular, the solution for a given right hand side is unique.

Proof. Define the operators $\mathcal{A}_0$ and $\mathcal{A}_1$ from $W^{1,m}(\Omega)$ to $W^{1,m}(\Omega)'$ by

$$\langle \mathcal{A}_0(u), v\rangle := \int_\Omega \nabla v \, A(x, \nabla u) \quad \text{and} \quad \langle \mathcal{A}_1(u), v\rangle := \int_\Omega v \, a(x,u)$$

so that the operator $\mathcal{A} := \mathcal{A}_0 + \mathcal{A}_1$ is the operator defined in (2.2.8).

Let u and v be fixed functions in $W^{1,m}(\Omega)$, and let $\rho \in (0,\infty)$ be arbitrary. Define

$$\Omega_\rho := \Big\{ \frac{|\nabla u - \nabla v|}{|\nabla u| + |\nabla v|} \geq \rho \Big\}.$$

Then, by (2.3.16),

$$\begin{aligned}\langle \mathcal{A}_0(u) - \mathcal{A}_0(v), u - v\rangle &\geq \nu \int_\Omega \Big(\frac{|\nabla u - \nabla v|}{|\nabla u| + |\nabla v|}\Big)^{2-m} |\nabla u - \nabla v|^m \\ &\geq \nu\rho^{2-m} \int_{\Omega_\rho} |\nabla u - \nabla v|^m \\ &\geq \nu\rho^{2-m}\Big(\int_\Omega |\nabla u - \nabla v|^m - \rho^m \int_\Omega (|\nabla u| + |\nabla v|)^m\Big).\end{aligned}$$

By (2.3.17), a similar estimate holds for $\mathcal{A}_1$. Hence, adding up, we find $c > 0$ such that

$$\begin{aligned}&\|u-v\|^m_{W^{1,m}(\Omega)} \\ &\quad \leq c\Big(\frac{1}{\rho^{2-m}} \langle \mathcal{A}(u) - \mathcal{A}(v), u - v\rangle + \rho^m\Big(\|u\|^m_{W^{1,m}(\Omega)} + \|v\|^m_{W^{1,m}(\Omega)}\Big)\Big)\end{aligned} \tag{2.3.19}$$

for all $\rho \in (0,\infty)$.

Now we set $v := 0$ in (2.3.19). Then we obtain that

$$\langle \mathcal{A}(u), u\rangle \geq \rho^{2-m}(c^{-1} - \rho^m)\|u\|^m_{W^{1,m}(\Omega)} - \|\mathcal{A}(0)\|_{W^{1,m}(\Omega)'}\|u\|_{W^{1,m}(\Omega)}.$$

Picking $\rho > 0$ so small that $\rho^m < c^{-1}$, we obtain that

$$\lim_{\|u\|_{W^{1,m}(\Omega)}\to\infty} \frac{\langle \mathcal{A}(u), u\rangle}{\|u\|_{W^{1,m}(\Omega)}} = \infty$$

since $m > 1$, i.e., $\mathcal{A}$ is coercive.

Since the remaining conditions on $\mathcal{A}$ in Theorem 1.5.5 can be checked as in the proof of Theorem 2.2.6, this proves that $\mathcal{A}$ is surjective. Hence (1.9.7) has a weak solution in $W^{1,m}(\Omega)$ for every right hand side, see Lemma 2.2.3.

Moreover, like in Theorem 2.2.6 we see that

$$\|u\|_{W^{1,m}(\Omega)} \leq c\Big(\|\mathcal{A}(u)\|^{1/(m-1)}_{W^{1,m}(\Omega)'} + 1\Big) \tag{2.3.20}$$

with a constant $c \geq 0$ that does not depend on u.

In order to show (2.3.18), let u and v be solutions for right hand sides (f_0, f_j, g) and $(\tilde{f}_0, \tilde{f}_j, \tilde{g})$, respectively. We pick $\rho \in (1, \infty)$ such that the two summands in (2.3.19) become equal, i.e.,

$$\rho^2 = \frac{\langle \mathcal{A}(u) - \mathcal{A}(v), u - v\rangle}{\|u\|^m_{W^{1,m}(\Omega)} + \|v\|^m_{W^{1,m}(\Omega)}}. \tag{2.3.21}$$

Note that if the numerator or the denominator in this definition equal zero, then $u = v$, and there is nothing to show. We obtain that

$$\begin{aligned}
&\|u - v\|^m_{W^{1,m}(\Omega)} \\
&\quad\leq 2c\,\Big(\langle \mathcal{A}(u) - \mathcal{A}(v), u - v\rangle\Big)^{m/2}\Big(\|u\|^m_{W^{1,m}(\Omega)} + \|v\|^m_{W^{1,m}(\Omega)}\Big)^{(2-m)/2} \\
&\quad\leq 2c\,\|\mathcal{A}(u) - \mathcal{A}(v)\|^{m/2}_{W^{1,m}(\Omega)'}\|u - v\|^{m/2}_{W^{1,m}(\Omega)}\Big(\|u\|^m_{W^{1,m}(\Omega)} + \|v\|^m_{W^{1,m}(\Omega)}\Big)^{(2-m)/2}
\end{aligned}$$

for a constant $c > 0$. Thus, using (2.3.20) and Lemma 1.1.3,

$$\begin{aligned}
&\|u - v\|_{W^{1,m}(\Omega)} \\
&\quad\leq c\,\Big(\|\mathcal{A}(u)\|_{W^{1,m}(\Omega)'} + \|\mathcal{A}(v)\|_{W^{1,m}(\Omega)'} + 1\Big)^{(2-m)/(m-1)}\|\mathcal{A}(u) - \mathcal{A}(v)\|_{W^{1,m}(\Omega)'}
\end{aligned}$$

for some constant $c \geq 0$. Since the norms of $\mathcal{A}(u)$, $\mathcal{A}(v)$, and $\mathcal{A}(u) - \mathcal{A}(v)$ can be estimated by the norms of (f_0, f_j, g), $(\tilde{f}_0, \tilde{f}_j, \tilde{g})$, and $(f_0 - \tilde{f}_0, f_j - \tilde{f}_j, g - \tilde{g})$, respectively, this implies (2.3.18). □

Example 2.3.23. If $m \in (1, 2]$ and $\omega > 0$. Then for the equation

$$\begin{cases} -\operatorname{div}\Big(|\nabla u|^{m-2}\nabla u\Big) + \omega |u|^{m-2} u = f_0 - \sum_{j=1}^{N} D_j f_j, & \text{on } \Omega, \\ |\nabla u|^{m-2}\dfrac{\partial u}{\partial \nu} = g + \sum_{j=1}^{N} f_j \nu_j, & \text{on } \partial\Omega \end{cases}$$

the estimate (2.3.18) holds.

Proof. The growth bounds in Assumption 2.3.21 are trivial for this example, and the monotonicity conditions (2.3.16) and (2.3.17) are satisfied by Lemma 1.1.5. □

3 Elliptic equations with Robin boundary conditions

In Chapter 2, we investigated existence, continuous dependence, and regularity for Neumann problems. In this chapter, we generalize the existence and regularity results to Robin problems, but we omit the investigation of continuous dependence in this setting because the ideas of Section 2.3 carry over to Robin boundary value problems without any difficulties if we assume in addition to some natural growth bound on the function h in (1.9.6) that h satisfies similar structural conditions as the coefficients, i.e., if we assume that h is linear in the setting of Section 2.3.1, if we assume that h is smooth with well-behaved derivatives in the approach of Section 2.3.2, and if we assume that h is non-decreasing in its second argument in the approach of Section 2.3.3.

For the existence results, even though the strategy to prove existence of solutions is the same as for Neumann problems, the pseudo-monotonicity of the associated operator is not obvious, which makes it worthwhile to look at the results for the Robin case in detail.

However, the main point of this chapter is to prove Hölder continuity of solutions for certain classes of Robin problems. The strategy here will not be the same as in Section 2.1. We rather reduce the question to the corresponding problem for a Neumann equation. For this we require that the corresponding Neumann problem is well-behaved, as it is the case for the examples of Chapter 2.

3.1 Existence of solutions

As in Section 2.2, the strategy to prove existence of solutions will be to show that the corresponding operator satisfies the assumptions of Theorem 1.5.5. To guarantee that the operator is coercive, we impose either structure conditions onto the Robin term h, or we have to shift the equation, i.e., consider

$$\begin{cases} -\operatorname{div} A(x,u,\nabla u) + a(x,u,\nabla u) + \omega|u|^{m-2}u = f_0 - \sum_{j=1}^{N} D_j f_j & \text{on } \Omega, \\ A(z,u,\nabla u)\cdot\nu + h(z,u) = g + \sum_{j=1}^{N} f_j\nu_j & \text{on } \partial\Omega \end{cases} \tag{3.1.1}$$

for sufficiently large ω.

For simplicity, we state the theorem only for $m < N$. The arguments for the cases $m = N$ and $m > N$ are similar, but the exponents for the Sobolev embeddings are different.

Theorem 3.1.1. *Let Ω be a Lipschitz domain and $m \in (1, N)$. In addition to Assumption 2.2.4, let*

$$|h(z,u)| \leq \psi_8(z)|u|^{m-1} + \psi_9(z) \tag{3.1.2}$$

for all $u \in \mathbb{R}$ and $z \in \partial\Omega$ with $\psi_8 \in L^{\frac{N-1}{m-1}+\varepsilon}(\partial\Omega)$ and $\psi_9 \in L^{\frac{(N-1)m}{(m-1)N}+\varepsilon}(\partial\Omega)$ for some $\varepsilon > 0$. Assume that h is continuous in its second argument. Then there exists ω_0 such that for every $\omega > \omega_0$ the equation (3.1.1) has a weak solution in $W^{1,m}(\Omega)$ for every right hand side in the spaces indicated in Lemma 2.2.3.

Proof. For $\omega \geq 0$, we define the operators $\mathcal{A}^\omega$ and $\mathcal{B}$ from $W^{1,m}(\Omega)$ to $W^{1,m}(\Omega)'$ by

$$\langle \mathcal{A}^\omega(u), v\rangle := \int_\Omega \nabla v\, A(x,u,\nabla u) + \int_\Omega v\, a(x,u,\nabla u) + \omega \int_\Omega v\, |u|^{m-2}u$$

and

$$\langle \mathcal{B}(u), v\rangle := \int_{\partial\Omega} v\, h(z,u)$$

and set $\mathcal{A}_R^\omega := \mathcal{A}^\omega + \mathcal{B}$. We claim that for sufficiently large ω the operator $\mathcal{A}_R^\omega$ satisfies the assumptions of Theorem 1.5.5.

As seen in the proof of Theorem 2.2.6, $\mathcal{A}^\omega$ is bounded for every ω. So in order to show that $\mathcal{A}_R^\omega$ is bounded, it suffices to show that $\mathcal{B}$ is bounded. For this we use (3.1.2) and Hölder's inequality to estimate

$$\begin{aligned}\left|\int_{\partial\Omega} v\, h(z,u)\right| &\leq \int_{\partial\Omega} |\psi_8|\, |u|^{m-1}|v| + \int_{\partial\Omega} |\psi_9|\, |v| \\ &\leq \|\psi_8\|_{L^{\frac{N-1}{m-1}}(\partial\Omega)} \left\||u|^{m-1}\right\|_{L^{\frac{(N-1)m}{(N-m)(m-1)}}(\partial\Omega)} \|v\|_{L^{\frac{(N-1)m}{N-m}}(\partial\Omega)} \\ &\quad + \|\psi_9\|_{L^{\frac{(N-1)m}{(m-1)N}}(\partial\Omega)} \|v\|_{L^{\frac{(N-1)m}{N-m}}(\partial\Omega)}.\end{aligned}$$

Thus from Theorem 1.3.10 we obtain that

$$\left|\int_{\partial\Omega} v\, h(z,u)\right| \leq \left(c_1 \|u\|_{W^{1,m}(\Omega)}^{m-1} + c_2\right) \|v\|_{W^{1,m}(\Omega)}$$

for some constants c_1 and c_2 depending only on the functions ψ_8 and ψ_9 and the operator norm of the trace operator. Thus $\mathcal{B}$ is bounded.

Let (u_n) be an arbitrary sequence in $W^{1,m}(\Omega)$ such that $u_n \rightharpoonup u$ in $W^{1,m}(\Omega)$ for some $u \in W^{1,m}(\Omega)$. Then $u_n \to u$ in $L^r(\partial\Omega)$ for every $r < \frac{(N-1)m}{N-m}$ because the trace operator is compact by Theorem 1.3.10. We can pick r so close to $\frac{(N-1)m}{N-m}$ that

$$\frac{r \cdot \frac{(N-1)m}{(m-1)N}}{r - (m-1)\frac{(N-1)m}{(m-1)N}} = \frac{(N-1)m}{(m-1)N - (N-m)(m-1)\frac{(N-1)m}{r(N-m)}} < \frac{N-1}{m-1} + \varepsilon.$$

Thus, by Lemma 1.4.3, $h(\cdot, u_n(\cdot))$ converges to $h(\cdot, u(\cdot))$ in $L^{\frac{(N-1)m}{(m-1)N}}(\partial\Omega)$. Since by Hölder's inequality and Theorem 1.3.10

$$\begin{aligned}\left|\langle \mathcal{B}(u_n) - \mathcal{B}(u), v\rangle\right| &\leq \|h(\cdot, u_n(\cdot)) - h(\cdot, u(\cdot))\|_{L^{\frac{(N-1)m}{(m-1)N}}(\partial\Omega)} \|v\|_{L^{\frac{(N-1)m}{N-m}}(\partial\Omega)} \\ &\leq c\,\|h(\cdot, u_n(\cdot)) - h(\cdot, u(\cdot))\|_{L^{\frac{(N-1)m}{(m-1)N}}(\partial\Omega)} \|v\|_{W^{1,m}(\Omega)}\end{aligned}$$

for some constant $c \geq 0$, this implies that $\mathcal{B}(u_n)$ converges to $\mathcal{B}(u)$ in $W^{1,m}(\Omega)'$. Hence $\mathcal{B}$ is completely continuous from $W^{1,m}(\Omega)$ to $W^{1,m}(\Omega)'$. Moreover, the operator $\mathcal{A}^\omega$ is pseudo-monotone by Theorem 2.2.5, compare also the proof of Theorem 2.2.6. Thus $\mathcal{A}^\omega_R$ is pseudo-monotone by Theorem 1.5.4.

Finally, we show that $\mathcal{A}^\omega_R$ is coercive. For every $u \in W^{1,m}(\Omega)$, we have by Hölder's inequality that

$$\begin{aligned}\left|\langle \mathcal{B}(u), u\rangle\right| &\leq \int_\Omega |\psi_8|\,|u|^m + \int_\Omega |\psi_9|\,|u| \\ &\leq \|\psi_8\|_{L^{r/(r-m)}(\partial\Omega)} \|u\|^m_{L^r(\partial\Omega)} + \|\psi_9\|_{L^{r'}(\partial\Omega)} \|u\|_{L^r(\partial\Omega)},\end{aligned}$$

where we pick $m < r < \frac{(N-1)m}{N-m}$ so close to $\frac{(N-1)m}{N-m}$ that ψ_8 is in $L^{r/(r-m)}(\partial\Omega)$ and ψ_9 is in $L^{r'}(\partial\Omega)$. Thus by Theorem 1.3.10 and Example 1.4.8, for every $\varepsilon > 0$ there exists $c_\varepsilon \geq 0$ such that

$$\left|\langle \mathcal{B}(u), u\rangle\right| \leq \varepsilon \|u\|^m_{W^{1,m}(\Omega)} + c_\varepsilon \|u\|^m_{L^m(\Omega)} + c_\varepsilon \|u\|_{W^{1,m}(\Omega)}$$

for all $u \in W^{1,m}(\Omega)$. Since by (2.2.13), there exist $\nu > 0$, $\omega_1 \geq 0$, and $d \geq 0$ such that

$$\langle \mathcal{A}(u), u\rangle \geq \nu \|u\|^m_{W^{1,m}(\Omega)} - \omega_1 \|u\|^m_{L^m(\Omega)} - d$$

for all $u \in W^{1,m}(\Omega)$, we obtain for $\varepsilon := \frac{\nu}{2}$ that

$$\begin{aligned}\langle \mathcal{A}^\omega_R(u), u\rangle &\geq \frac{\nu}{2}\|u\|^m_{W^{1,m}(\Omega)} + (\omega - \omega_1 - c_\varepsilon)\|u\|^m_{L^m(\Omega)} - d - c_\varepsilon \|u\|_{W^{1,m}(\Omega)} \\ &\geq \frac{\nu}{2}\|u\|^m_{W^{1,m}(\Omega)} - d - c_\varepsilon \|u\|_{W^{1,m}(\Omega)}\end{aligned}$$

for $\omega > \omega_0 := \omega_1 + c_\varepsilon$. This proves that $\mathcal{A}^\omega_R$ is coercive for $\omega > \omega_0$.

We have checked that $\mathcal{A}^\omega_R$ satisfies the assumptions of Theorem 1.5.5 for $\omega \geq \omega_0$. Now the claim follows as in the proof of Theorem 2.2.6. □

The following observation is a consequence of the proof of Theorem 3.1.1.

Corollary 3.1.2. *In the situation of Theorem 3.1.1, assume that*

$$\exists M, L \geq 0\ \exists \varepsilon > 0 \text{ s.t. } h(z, u)u \geq -M|u|^{m-\varepsilon} - L \quad \forall z \in \partial\Omega\ \forall u \in \mathbb{R}, \tag{3.1.3}$$

for example, $h(z, u) \cdot u \geq 0$ *for all* $z \in \partial\Omega$ *and* $u \in \mathbb{R}$. *If* $\omega_0 \in \mathbb{R}$ *is as in* (2.2.13), *then* (3.1.1) *has a weak solution for every* $\omega \geq \omega_0$.

Proof. We use the same notation as in the proof of Theorem 3.1.1. Condition (3.1.3) implies that

$$\langle \mathcal{B}(u), u\rangle \geq -M\int_{\partial\Omega}|u|^{m-\varepsilon} - \int_{\partial\Omega} L \geq -M'\|u\|^{m-\varepsilon}_{W^{1,m}(\Omega)} - L' \tag{3.1.4}$$

for some constants $M', L' \geq 0$ by Theorem 1.3.10. Hence

$$\begin{aligned}\langle \mathcal{A}^\omega_R(u), u\rangle &= \langle \mathcal{A}^\omega(u), u\rangle + \langle \mathcal{B}(u), u\rangle \\ &\geq \nu'\|u\|^m_{W^{1,m}(\Omega)} + (\omega-\omega_0)\|u\|^m_{L^m(\Omega)} - d - M'\|u\|^{m-\varepsilon}_{W^{1,m}(\Omega)} - L'\end{aligned}$$

by (2.2.13) and (3.1.4), which proves that $\mathcal{A}^\omega_R$ is coercive for $\omega > \omega_0$. Since the coercivity was the only point in the proof of Theorem 3.1.1 for which we needed that ω be large, we can conclude as before that (3.1.1) admits a weak solution if $\omega > \omega_0$. □

Linear equations will be of particular interest for Chapter 4, and we treat them now separately. If the coefficients are bounded, the following proposition is a special case of Theorem 3.1.1. However, in order to allow unbounded coefficients, we now follow the standard approach and use the Lax-Milgram theorem.

Proposition 3.1.3. *Let Ω be a bounded Lipschitz domain. Under Assumption 1.9.3, let in addition β be a function in $L^{q-1}(\partial\Omega)$ for some $q > N$. Then there exists $\omega_0 > 0$ such that for all $\omega > \omega_0$ the equation*

$$\begin{cases} -\sum_{j=1}^N D_j\Big(\sum_{i=1}^N a_{ij}D_iu + b_ju\Big) + \Big(\sum_{i=1}^N c_iD_iu + du\Big) + \omega u = f_0 - \sum_{j=1}^N D_jf_j & \text{on } \Omega, \\ \sum_{j=1}^N\Big(\sum_{i=1}^N a_{ij}D_iu + b_ju\Big)\nu_j + \beta u = g + \sum_{j=1}^N f_j\nu_j & \text{on } \partial\Omega, \end{cases} \tag{3.1.5}$$

has a weak solution $u \in W^{1,2}(\Omega)$ whenever the functions f_0, f_j, and g are as in Proposition 2.2.2.

Proof. For u and v in $W^{1,2}(\Omega)$ define

$$\mathfrak{a}^\omega_\beta(u,v) := \mathfrak{a}^\omega(u,v) + \int_{\partial\Omega}\beta uv$$

with $\mathfrak{a}^\omega$ defined in (2.2.6). By Hölder's inequality,

$$\left|\int_{\partial\Omega}\beta uv\right| \leq \|\beta\|_{L^{q-1}(\partial\Omega)}\,\|u\|_{L^{2(q-1)/(q-2)}(\partial\Omega)}\,\|v\|_{L^{2(q-1)/(q-2)}(\partial\Omega)}$$

for all u and v in $W^{1,2}(\Omega)$. Since $q > N$, the trace operator is compact from $W^{1,2}(\Omega)$ to $L^{2(q-1)/(q-2)}(\partial\Omega)$ by Theorem 1.3.10. Now we obtain from Example 1.4.9 that for every

$\varepsilon > 0$ there exists $c_\varepsilon \geq 0$ such that

$$\left|\int_{\partial\Omega} \beta uv\right| \leq \|\beta\|_{L^{q-1}(\partial\Omega)} \Big(\varepsilon \|\nabla u\|_{L^2(\Omega)} + c_\varepsilon \|u\|_{L^2(\Omega)}\Big)\Big(\varepsilon \|\nabla v\|_{L^2(\Omega)} + c_\varepsilon \|v\|_{L^2(\Omega)}\Big).$$

Since $\mathfrak{a}^\omega$ is continuous, see the proof of Proposition 2.2.2, this shows that $\mathfrak{a}^\omega_\beta$ is a continuous form.

Moreover, we deduce using Lemma 2.2.1 and the estimate $(a+b)^2 \leq 2a^2 + 2b^2$, that there exist $\alpha > 0$ and $\omega_0 \in \mathbb{R}$ such that

$$\begin{aligned}\mathfrak{a}^\omega_\beta(u,u) \geq \alpha \|\nabla u\|^2_{L^2(\Omega)} + (\omega - \omega_0) \|u\|^2_{L^2(\Omega)} \\ - 2\|\beta\|_{L^{q-1}(\partial\Omega)} \Big(\varepsilon^2 \|\nabla u\|^2_{L^2(\Omega)} + c_\varepsilon^2 \|u\|^2_{L^2(\Omega)}\Big).\end{aligned}$$

Now, if we pick $\varepsilon > 0$ small enough, we obtain that there exist $\alpha' > 0$ and $\omega_0' \in \mathbb{R}$ such that

$$\mathfrak{a}^\omega_\beta(u,u) \geq \alpha' \|\nabla u\|^2_{L^2(\Omega)} + (\omega - \omega_0') \|u\|^2_{L^2(\Omega)}$$

for all $u \in W^{1,2}(\Omega)$. Thus, by Theorem 1.4.6, if $\omega > \omega_0'$, then for every $\varphi \in W^{1,2}(\Omega)'$ there exists a unique $u \in W^{1,2}(\Omega)$ such that $\mathfrak{a}^\omega_\beta(u,v) = \varphi(v)$ for all $v \in W^{1,2}(\Omega)$. Picking φ as in the proof of Proposition 2.2.2, we obtain a weak solution of (3.1.5). □

3.2 Hölder regularity

In this section we exhibit conditions under which the solutions of Robin problems are Hölder continuous up to the boundary. Our approach is based on the following simple key observation: if $u \in W^{1,m}(\Omega)$ is a weak solution of the Robin problem (1.9.6), then for every $\omega \in \mathbb{R}$ it is as well a weak solution of the Neumann problem

$$\begin{cases} -\operatorname{div} A(x,u,\nabla u) + a(x,u,\nabla u) + \omega|u|^{m-2}u = f_0 - \sum_{j=1}^N D_j f_j + \omega|u|^{m-2}u & \text{on } \Omega, \\ A(z,u,\nabla u)\cdot\nu = g + \sum_{j=1}^N f_j\nu_j - h(z,u) & \text{on } \partial\Omega. \end{cases} \tag{3.2.1}$$

Thus we can expect that results about (inhomogeneous) Neumann problems translate into results about Robin problems. The main difficulty is of course to check that the conditions of Corollary 2.1.6 are satisfied.

If h is bounded, then we can set $\omega := 0$ and apply Corollary 2.1.6. In fact, this still works if h grows very slowly as u tends to infinity. However, under the typical growth conditions, for example if we want to allow $h(z,u) = |u|^{m-2}u$ in the m-Laplace equation, the situation is more difficult. We apply bootstrapping to overcome this problem.

We treat here all the cases of Section 2.3 at once. The only important prerequisite is that for the corresponding Neumann problem there exists a solution for every right hand side and that the resolvent is Hölder continuous. We formulate this as Assumption 3.2.1. Note that we have already exhibited several examples where this assumption is fulfilled.

Assumption 3.2.1. *Let Ω be a bounded Lipschitz domain and let $m \in (1, N)$. Let the functions A and a be such that for some $\omega \in \mathbb{R}$ problem* (2.2.10) *has a weak solution in $W^{1,m}(\Omega)$ whenever the right hand side is in the spaces indicated in Lemma 2.2.3, and let this ω now be fixed.*

Assume that there exist $\beta \in (0, 1]$ and a constant $c \geq 0$ such that

$$\|u - \tilde{u}\|_{W^{1,m}(\Omega)} \leq c\Big(\|f_0 - \tilde{f}_0\| + \sum_{j=1}^{N} \|f_j - \tilde{f}_j\| + \|g - \tilde{g}\|\Big)^{\beta} \tag{3.2.2}$$

whenever u and $\tilde{u}$ are weak solutions of (2.2.10) *in $W^{1,m}(\Omega)$ for the right hand sides (f_0, f_j, g) and $(\tilde{f}_0, \tilde{f}_j, \tilde{g})$, respectively, where the norms are in the spaces indicated in Lemma 2.2.3; then in particular the solutions of problem* (2.2.10) *are unique.*

Finally, assume that for every $\varepsilon > 0$ there exist $\alpha \in (0, 1)$ and $c \geq 0$ such that the solution u is in $C^{0,\alpha}(\Omega)$ whenever $f_0 \in L^{N/m+\varepsilon}(\Omega)$, $f_j \in L^{N/(m-1)+\varepsilon}(\Omega)$, and $g \in L^{(N-1)/(m-1)+\varepsilon}(\partial\Omega)$ and satisfies the estimate

$$\|u\|_{C^{0,\alpha}(\Omega)} \leq c\Big(\Big(\|f_0\| + \sum_{j=1}^{N} \|f_j\| + \|g\|\Big)^{\beta} + 1\Big) \tag{3.2.3}$$

for the respective norms.

Remark 3.2.2. It is an essential feature of Assumption 3.2.1 that we have the same exponent β in (3.2.2) and (3.2.3). A particular case of the instance that these exponents differ will be discussed in the second half of this section, see Assumption 3.2.11.

We could allow that the number c in (3.2.2) and (3.2.3) depends on an upper bound for the functions in the spaces of Lemma 2.2.3. In fact, this would make no difference in the proof. But since this is not required for the situations we can treat with this set of assumptions, there is no need to include this additional notational difficulty.

Remark 3.2.3. Typically, we have $\beta = \frac{1}{m-1}$, see the examples in Chapter 2. Thus if $m < 2$, then the operator that maps a right hand side to the corresponding solution of (2.2.10) is constant by (3.2.2), which is absurd. Hence we expect Assumption 3.2.1 to hold only for $m \geq 2$. The case $m \in (1, 2]$ will be treated separately at the end of this section.

Lemma 3.2.4. *Let $\Omega \subset \mathbb{R}^N$ be a bounded Lipschitz domain, let $1 < m < N$, and let*

$\theta \in (0,1)$. *Define* r_1, r_2, *and* r_3 *by*

$$r_1 := \frac{mN}{\theta(mN+m-N)+(1-\theta)m^2}, \qquad r_2 := \frac{mN}{\theta N(m-1)+(1-\theta)m(m-1)},$$
$$r_3 := \frac{m(N-1)}{\theta N(m-1)+(1-\theta)m(m-1)}.$$

Under Assumption 3.2.1, if f_0 *is in* $L^{r_1+\varepsilon}(\Omega)$, *each* f_j *is in* $L^{r_2+\varepsilon}(\Omega)$, *and* g *is in* $L^{r_3+\varepsilon}(\partial\Omega)$ *for some* $\varepsilon > 0$, *then the unique weak solution* u *in* $W^{1,m}(\Omega)$ *of* (2.2.10) *satisfies*

$$u \in L^{\frac{Nm}{\theta(N-m)}}(\Omega) \quad \text{and} \quad u|_{\partial\Omega} \in L^{\frac{(N-1)m}{\theta(N-m)}}(\partial\Omega).$$

Proof. Since $r_1 \geq \frac{mN}{mN+m-N}$, $r_2 \geq \frac{m}{m-1}$, and $r_3 \geq \frac{m(N-1)}{mN-N}$, by assumption there exists a unique weak solution $u \in W^{1,m}(\Omega)$ whenever the right hand side is in the indicated spaces.

Let R be the (non-linear) operator assigning to a given right hand side the corresponding weak solution of (2.2.10). By Assumption 3.2.1, R is Hölder continuous from

$$A_1 := L^{mN/(mN+n-N)}(\Omega) \oplus L^{m/(m-1)}(\Omega)^N \oplus L^{m(N-1)/N(m-1)}(\partial\Omega)$$

to

$$W^{1,m}(\Omega) \hookrightarrow B_1 := L^{mN/(N-m)}(\Omega) \oplus L^{m(N-1)/(N-m)}(\partial\Omega)$$

with exponent β.

Moreover, by Assumption 3.2.1, for

$$A_0 := L^{N/m+\tilde{\varepsilon}}(\Omega) \oplus L^{N/(m-1)+\tilde{\varepsilon}}(\Omega)^N \oplus L^{(N-1)/(m-1)+\tilde{\varepsilon}}(\partial\Omega)$$

and

$$B_0 := L^\infty(\Omega) \oplus L^\infty(\partial\Omega)$$

with $\tilde{\varepsilon} > 0$ being arbitrary we have

$$\|R(f_0, f_j, g)\|_{B_0} \leq c_1 \|R(f_0, f_j, g)\|_{C^{0,\alpha}(\Omega)} \leq c_2\Big(\big(\|(f_0, f_j, g)\|_{A_0}\big)^\beta + 1\Big)$$

for constants $c_1 \geq 0$ and $c_2 \geq 0$.

To prove the continuity of R from A_0 to B_0, let (f_0, f_j, g) and $(\tilde{f}_0, \tilde{f}_j, \tilde{g})$ be two arbitrary right hand sides in A_0 and set $(u, u|_{\partial\Omega}) := R(f_0, f_j, g)$ and $(\tilde{u}, \tilde{u}|_{\partial\Omega}) := R(\tilde{f}_0, \tilde{f}_j, \tilde{g})$. By Theorem 1.6.1 there exist $c_0 \geq 0$ and $\eta \in (0,1)$ such that

$$\begin{aligned}\|(u, u|_{\partial\Omega}) - (\tilde{u}, \tilde{u}|_{\partial\Omega})\|_{B_0} &= \|u - \tilde{u}\|_{L^\infty(\Omega)} \leq c_0 \, \|u - \tilde{u}\|^\eta_{C^{0,\alpha}(\Omega)} \|u - \tilde{u}\|^{1-\eta}_{L^m(\Omega)} \\ &\leq c_0 \, c^\eta \Big(\|(f_0, f_j, g)\|^\beta_{A_0} + \|(\tilde{f}_0, \tilde{f}_j, \tilde{g})\|^\beta_{A_0} + 2\Big)^\eta c^{1-\eta} \|(f_0, f_j, g) - (\tilde{f}_0, \tilde{f}_j, \tilde{g})\|^{(1-\eta)\beta}_{A_1}.\end{aligned}$$

Since the norm in A_1 can be estimated by the norm in A_0, this shows that R is locally Hölder continuous from A_0 to B_0.

Now the claim follows from Theorem 1.6.12. □

The following theorem is the main regularity result for Robin problems. We have seen in Chapter 2 that large classes of equations satisfy its assumptions, and we will point this out once again in the examples. The complementary existence results are contained in the previous section. Nevertheless, they are also repeated in the examples.

Theorem 3.2.5. *Let $\Omega \subset \mathbb{R}^N$ be a bounded Lipschitz domain and $1 < m < N$. Let h be a measurable function, and assume that there exist $\varepsilon > 0$ and functions ψ_8 and ψ_9 in $L^{(N-1)/(m-1)+\varepsilon}(\partial\Omega)$ such that*

$$|h(z,u)| \leq \psi_8(z)|u|^{m-1} + \psi_9(z) \tag{3.2.4}$$

for all $z \in \partial\Omega$ and all $u \in \mathbb{R}$. Under Assumption 3.2.1, there exists $\alpha \in (0,1)$ such that every weak solution $u \in W^{1,m}(\Omega)$ of the Robin problem (1.9.6) is in $C^{0,\alpha}(\Omega)$ whenever f_0 is in $L^{N/m+\varepsilon}(\Omega)$, each f_j is in $L^{N/(m-1)+\varepsilon}(\Omega)$, and g is in $L^{(N-1)/(m-1)+\varepsilon}(\partial\Omega)$ for some $\varepsilon > 0$.

Proof. Let $u \in W^{1,m}(\Omega)$ be a weak solution of (1.9.6). For $\theta \in [0,1]$, define

$$p_\theta := \frac{mN}{\theta(mN+m-N)+(1-\theta)m^2} = \frac{mN}{m^2+\theta(N-m)(m-1)} \in (1,\infty),$$
$$q_\theta := \frac{m(N-1)}{\theta N(m-1)+(1-\theta)m(m-1)} = \frac{m(N-1)}{m(m-1)+\theta(N-m)(m-1)} \in (1,\infty).$$

We are going to show that $|u|^{m-1} \in L^p(\Omega)$ and $h(\cdot,u(\cdot)) \in L^q(\partial\Omega)$ for some $p > p_0 = \frac{N}{m}$ and $q > q_0 = \frac{N-1}{m-1}$. In other words, we show that the set

$$\Theta := \left\{\theta \in [0,1] : \exists \varepsilon > 0 \text{ such that } |u|^{m-1} \in L^{p_\theta+\varepsilon}(\Omega) \text{ and } h(\cdot,u(\cdot)) \in L^{q_\theta+\varepsilon}(\partial\Omega)\right\}.$$

contains 0.

Since p_θ and q_θ are decreasing functions in θ, for $0 \leq \theta_1 \leq \theta_2 \leq 1$ we have that $\theta_1 \in \Theta$ implies $\theta_2 \in \Theta$. Moreover, since p_θ and q_θ depend continuously on θ, Θ is a relatively open subset of $[0,1]$. Thus either $\Theta = \emptyset$, $\Theta = (a,1]$ for some $a \in [0,1)$, or $\Theta = [0,1]$. Therefore it suffices to rule out the first two possibilities in order to conclude that $0 \in \Theta$.

By Theorem 1.3.9 we have that $u \in L^{mN/(N-m)}(\Omega)$. Since

$$\frac{mN}{(N-m)(m-1)} = \frac{mN}{mN+m-N-m^2} > \frac{mN}{mN+m-N} = p_1,$$

this implies that

$$|u|^{m-1} \in L^{\frac{mN}{(N-m)(m-1)}}(\Omega) \hookrightarrow L^{p_1+\varepsilon}(\Omega)$$

for some $\varepsilon > 0$. Moreover, Theorem 1.3.10 asserts that $u|_{\partial\Omega}$ is in $L^{m(N-1)/(N-m)}(\partial\Omega)$. Hence the trace of $|u|^{m-1}$ is in $L^{\frac{m(N-1)}{(m-1)(N-m)}}(\partial\Omega)$, where we note that

$$\frac{m(N-1)}{(m-1)(N-m)} \geq \frac{m}{m-1} > 1.$$

Hence Hölder's inequality asserts that $\psi_8|u|^{m-1}$ is in $L^{r_1}(\partial\Omega)$ for $r_1 \in (1,\infty)$ defined by

$$\frac{1}{r_1} = \frac{(m-1)(N-m)}{m(N-1)} + \frac{m-1-\varepsilon}{N-1} < \frac{(m-1)(N-m)+m(m-1)}{m(N-1)} = \frac{N(m-1)}{m(N-1)}$$

for some $\varepsilon > 0$. Thus $r_1 > q_1$. Since also $\frac{N-1}{m-1} + \varepsilon > q_1$, we deduce from (3.2.4) that $h(\cdot, u(\cdot)) \in L^{q_1+\varepsilon}(\partial\Omega)$ for some $\varepsilon > 0$. We have checked that $1 \in \Theta$, i.e., $\Theta \neq \emptyset$.

Assume that $\Theta = (a, 1]$ for some $a \in [0, 1)$. Let $\theta \in (a, 1]$ be arbitrary, and let ω be as in Assumption 3.2.1. Then the right hand side of (3.2.1) satisfies the assumptions of Lemma 3.2.4 for this θ. In fact, f_0, f_j, and g satisfy these assumptions even for $\theta = 0$, thus for all $\theta \in [0, 1]$, and for the terms involving u this is the definition of θ being an element of Θ. Thus we obtain from Lemma 3.2.4 that

$$u \in L^{\frac{mN}{\theta(N-m)}}(\Omega) \quad \text{and} \quad u|_{\partial\Omega} \in L^{\frac{(N-1)m}{\theta(N-m)}}(\partial\Omega)$$

and thus

$$|u|^{m-1} \in L^{\frac{mN}{\theta(N-m)(m-1)}}(\Omega)$$

for all $\theta > a$. Since

$$\frac{mN}{a(N-m)(m-1)} > \frac{mN}{m^2 + a(N-m)(m-1)} = p_a,$$

by the continuity of these expressions there exists $\theta_0 > a$, i.e., $\theta_0 \in \Theta$, such that

$$\frac{mN}{\theta_0(N-m)(m-1)} > p_a.$$

Hence $|u|^{m-1} \in L^{p_a+\varepsilon}(\Omega)$ for some $\varepsilon > 0$. Moreover, using Hölder's inequality in the same way as before, we obtain that $\psi_8|u|^{m-1}$ is in $L^{r_\theta}(\partial\Omega)$ with $r_\theta > 1$ given by

$$\begin{aligned}\frac{1}{r_\theta} &= \frac{\theta(N-m)(m-1)}{m(N-1)} + \frac{m-1-\varepsilon}{N-1} \\ &= \frac{m(m-1-\varepsilon)+\theta(N-m)(m-1)}{m(N-1)} = \frac{1}{q_\theta} - \frac{\varepsilon}{N-1},\end{aligned}$$

for some $\varepsilon > 0$. Exploiting the continuity of these expressions, we see that there exists $\theta_1 > a$, i.e., $\theta_1 \in \Theta$, such that $\frac{1}{r_{\theta_1}} < \frac{1}{q_a} - \varepsilon$ for some $\varepsilon > 0$. Hence $r_{\theta_1} = q_a + \varepsilon$ for some $\varepsilon > 0$. From (3.2.4) we thus obtain that $h(\cdot, u(\cdot)) \in L^{q_a+\varepsilon}(\partial\Omega)$ for some $\varepsilon > 0$. Hence $a \in \Theta$, contradicting the assumption $\Theta = (a, 1]$.

Since neither $\Theta = \emptyset$ nor $\Theta = (a, 1]$, the introductory comments about Θ imply that $\Theta = [0, 1]$, i.e., $0 \in \Theta$. This means that u is in $L^{N/m+\varepsilon}(\Omega)$ and that $h(\cdot, u(\cdot))$ is in $L^{(N-1)/(m-1)+\varepsilon}(\partial\Omega)$ for some $\varepsilon > 0$. Thus by Assumption 3.2.1, the right hand side of (3.2.1) is regular enough to ensure that we can apply Corollary 2.1.6 in order to conclude that u is in $C^{0,\alpha}(\Omega)$ for some $\alpha \in (0, 1)$. □

Remark 3.2.6. The statement of the theorem is restricted to $m < N$ because the case $m \geq N$ is trivial. In fact, if $m > N$, then Hölder continuity of functions in $W^{1,m}(\Omega)$ follows from the Sobolev embedding theorem. In the border case $m = N$ we have $u \in L^r(\Omega)$ and $u|_{\partial\Omega} \in L^r(\partial\Omega)$ for every $r < \infty$, thus $h(\cdot, u) \in L^{(N-1)/(m-1)+\varepsilon}(\partial\Omega)$ for some $\varepsilon > 0$ by (3.2.4). Hence we can directly apply Corollary 2.1.6 to (3.2.1) to obtain the Hölder continuity of solutions. In fact, for $m \geq N$ we thus need no information about existence or continuous dependence for the Neumann problem.

In Chapter 2 we have found sufficient conditions under which Assumption 3.2.1 is fulfilled. Since all necessary ingredients can be found there, we content ourselves with repeating the conclusion for our prime examples. It is of course easy to come up with modifications of these situations to which the results of Chapter 2 still apply, for example adding lower order terms, letting the coefficients in Examples 3.2.9 and 3.2.10 depend on x, or replacing arctan in Example 3.2.9 by some other smooth, strictly positive, increasing, bounded function.

Example 3.2.7. Let Ω be a bounded Lipschitz domain. Under Assumption 1.9.3, let in addition β be a function in $L^{q-1}(\partial\Omega)$ for $q > N$. If u is a weak solution in $W^{1,2}(\Omega)$ of (3.1.5), which by Proposition 3.1.3 exists for sufficiently large ω, and if $f_0 \in L^{q/2}(\Omega)$, $f_j \in L^q(\Omega)$, and $g \in L^{q-1}(\partial\Omega)$ with $q > N$, then u is Hölder continuous.

Proof. We check Assumption 3.2.1 for $m = 2$ and $\beta = 1$, where the coefficients for the quasi-linear problem and the coefficients for the linear problem are related by (1.9.5). The existence of solutions of (2.2.10) follows from Proposition 2.2.2. The Hölder continuous dependence in the norm of $W^{1,2}(\Omega)$ of those solutions is Proposition 2.3.1. Finally, the Hölder regularity including the norm bound, follows from Corollary 2.1.6 since the norm of the solution in $L^2(\Omega)$ occurring in (2.1.6) can be estimated by Proposition 2.2.2. Thus Assumption 3.2.1 is satisfied. Now the claim follows from Theorem 3.2.5. □

Remark 3.2.8. A more general version of the result in Example 3.2.7, including mixed boundary conditions, certain systems of equations, and continuous dependence on the coefficients, has been proved by Griepentrog and Recke [GR01].

Example 3.2.9. Let $\Omega \subset \mathbb{R}^N$ be a bounded Lipschitz domain, $N \geq 3$, and $\omega \geq 0$. Let h be measurable and satisfy

$$|h(z, u)| \leq \psi_8(z)|u| + \psi_9(z)$$

with functions ψ_8 and ψ_9 in $L^{q-1}(\partial\Omega)$ for some $q > N$, for example $h(z, u) = \beta u$ with $\beta \in L^\infty(\Omega)$. Let $f_0 \in L^{q/2}(\Omega)$, $f_j \in L^q(\Omega)$, and $g \in L^{q-1}(\partial\Omega)$ for some $q > N$. If u is a weak solution $u \in W^{1,2}(\Omega)$ of

$$\begin{cases} -\operatorname{div}\Big(\big(1+\arctan(|\nabla u|^2)\big)\nabla u\Big) + \omega u^2 = f_0 - \sum_{j=1}^{N} f_j & \text{on } \Omega, \\ \big(1+\arctan(|\nabla u|^2)\big)\dfrac{\partial u}{\partial \nu} + h(z,u) = g + \sum_{j=1}^{N} f_j \nu_j & \text{on } \partial\Omega, \end{cases}$$

then u is Hölder continuous. Moreover, if h is continuous in its second argument and if ω is large enough, then for every right hand side there exists a weak solution. If in addition h satisfies (3.1.3), then every $\omega > 0$ is "large enough". Finally, if h is non-decreasing in its second argument, then the solution is unique for every $\omega > 0$.

Proof. We check Assumption 3.2.1 for $m = 2$ and $\beta = 1$ with

$$A(x,u,p) = \big(1+\arctan(|p|^2)\big)p^T \text{ and } a(x,u,p) = 0.$$

The existence of solutions of the Neumann problem (2.2.10) follows from Example 2.2.9. The Hölder continuous dependence in the norm of $W^{1,2}(\Omega)$ of those solutions is the statement of Example 2.3.7. Finally, the Hölder regularity including the norm bound follows from Corollary 2.1.6 since the norm of the solution in $L^2(\Omega)$ occurring in (2.1.6) can be estimated using (2.2.11). Thus Assumption 3.2.1 is satisfied. Now Hölder continuity follows from Theorem 3.2.5. The next two claims are the statements of Theorem 3.1.1 and Corollary 3.1.2, compare also Example 2.2.9 for the fact that (2.2.13) is satisfied with $\omega_0 = 0$. Finally, if h is non-decreasing in its second argument, then the operator $\mathcal{B}$ is monotone. Hence the operator $\mathcal{A}^{\omega}_{R}$, defined as in the proof of Theorem 3.1.1, is strictly monotone for $\omega > 0$ in the sense of Definition 1.5.1. As in Corollary 2.2.7 we can thus conclude that the solution is unique. □

Example 3.2.10 (m-Laplace equation, $m \geq 2$)**.** Let $\Omega \subset \mathbb{R}^N$ be a bounded Lipschitz domain, let $2 \leq m < N$, and let $\omega \geq 0$. Let h be measurable and satisfy

$$|h(z,u)| \leq \psi_8 |u|^{m-1} + \psi_9$$

with function ψ_8 and ψ_9 in $L^{(N-1)/(m-1)+\varepsilon}(\partial\Omega)$ for some $\varepsilon > 0$, e.g., $h(z,u) = \beta|u|^{m-1}$ with $\beta \in L^\infty(\Omega)$. Assume $f_0 \in L^{N/m+\varepsilon}(\Omega)$, $f_j \in L^{N/(m-1)+\varepsilon}(\Omega)$, and $g \in L^{(N-1)/(m-1)+\varepsilon}(\partial\Omega)$ for some $\varepsilon > 0$, and assume that $u \in W^{1,m}(\Omega)$ is a weak solution of

$$\begin{cases} -\operatorname{div}\big(|\nabla u|^{m-2}\nabla u\big) + \omega |u|^{m-2}u = f_0 - \sum_{j=1}^{N} f_j, & \text{on } \Omega, \\ |\nabla u|^{m-2}\dfrac{\partial u}{\partial \nu} + h(z,u) = g + \sum_{j=1}^{N} f_j \nu_j, & \text{on } \partial\Omega. \end{cases}$$

Then u is Hölder continuous. Moreover, if h is continuous in its second argument and if ω is large enough, then for every right hand side there exists a weak solution. If in addition

h satisfies (3.1.3), then every $\omega > 0$ is "large enough". Finally, if h is non-decreasing in its second argument, then the solution is unique.

Proof. We check Assumption 3.2.1 for $\beta = \frac{1}{m-1}$ with

$$A(x,u,p) = |p|^{m-2}p^T \text{ and } a(x,u,p) = 0.$$

The existence of solutions of the Neumann problem (2.2.10) is a special case of Example 2.2.10. The Hölder continuous dependence in the norm of $W^{1,m}(\Omega)$ of those solutions is the statement of Example 2.3.20. Finally, the Hölder regularity including the norm bound follows from Corollary 2.1.6 since the norm of the solution on $L^m(\Omega)$ occurring in (2.1.6) can be estimated using (2.2.11). Thus Assumption 3.2.1 is satisfied. Now Hölder regularity follows from Theorem 3.2.5. The additional claims follow as in Example 3.2.9, compare also Example 2.2.10 to see that (2.2.13) holds with $\omega_0 = 0$. □

Theorem 3.2.5 covers many interesting examples. However, the m-Laplace equation does not satisfy Assumption (3.2.1) if $m \in (1,2)$, see Remark 3.2.3. The aim of the rest of this section is to overcome this problem for the whole class of problems covered by Theorem 2.3.22, so in particular for the m-Laplace equation, Example 2.3.23. In fact, in this situation we do have estimates of the kind required in Assumption 3.2.1, but for different exponents. This allows us to proceed in a similar manner as before.

As before, we formulate in an assumption what we need to know about the equation. To keep things reasonably simple, we specialize the exponents in this assumptions to the typical ones, i.e., to those obtained by Corollary 2.1.6 and Theorem 2.3.22.

Assumption 3.2.11. *Let Ω be a bounded Lipschitz domain and let $m \in (1,N)$. Let the functions A and a be such that for some $\omega \in \mathbb{R}$ problem* (2.2.10) *has a weak solution in $W^{1,m}(\Omega)$ whenever the right hand side is in the spaces indicated in Lemma 2.2.3.*

Moreover, assume that

$$\|u - \tilde{u}\|_{W^{1,m}(\Omega)} \le c\Big(\|f_0 - \tilde{f}_0\| + \sum_{j=1}^{N} \|f_j - \tilde{f}_j\| + \|g - \tilde{g}\|\Big) \tag{3.2.5}$$

whenever u and $\tilde{u}$ are weak solutions in $W^{1,m}(\Omega)$ for the right hand sides (f_0, f_j, g) and $(\tilde{f}_0, \tilde{f}_j, \tilde{g})$, respectively, where the norms are in the spaces indicated in Lemma 2.2.3; then in particular the solutions of problem (2.2.10) *are unique. Here, $c \ge 0$ is allowed to depend on upper bounds of the norms of (f_0, f_j, g) and $(\tilde{f}_0, \tilde{f}_j, \tilde{g})$ in those spaces.*

Finally, assume that for every $\varepsilon > 0$ there exist $\alpha \in (0,1)$ and $c \ge 0$ such that the solution u is in $C^{0,\alpha}(\Omega)$ whenever $f_0 \in L^{N/m+\varepsilon}(\Omega)$, $f_j \in L^{N/(m-1)+\varepsilon}(\Omega)$, and $g \in L^{(N-1)/(m-1)+\varepsilon}(\partial\Omega)$ and satisfies the estimate

$$\|u\|_{C^{0,\alpha}(\Omega)} \le c\bigg(\Big(\|f_0\| + \sum_{j=1}^{N} \|f_j\| + \|g\|\Big)^{1/(m-1)} + 1\bigg) \tag{3.2.6}$$

for the respective norms.

As before, we could also allow that the value of c in (3.2.6) depends on upper bounds for (f_0, f_j, g) in the spaces of Lemma 2.2.3, but this is not needed for our examples.

The following lemma is proved like Lemma 3.2.4.

Lemma 3.2.12. *Let $\Omega \subset \mathbb{R}^N$ be a bounded Lipschitz domain, let $1 < m < N$, let $\theta \in (0,1)$, and define $\eta \in (0,1)$ by*

$$\frac{1-\eta}{\eta} = \frac{(m-1)(1-\theta)}{\theta}. \tag{3.2.7}$$

Define r_1, r_2, and r_3 by

$$r_1 := \frac{mN}{\theta(mN+m-N)+(1-\theta)m^2}, \qquad r_2 := \frac{mN}{\theta N(m-1)+(1-\theta)m(m-1)},$$
$$r_3 := \frac{m(N-1)}{\theta N(m-1)+(1-\theta)m(m-1)}.$$

Under Assumption 3.2.11, if f_0 is in $L^{r_1+\varepsilon}(\Omega)$, each f_j is in $L^{r_2+\varepsilon}(\Omega)$, and g is in $L^{r_3+\varepsilon}(\partial\Omega)$ for some $\varepsilon > 0$, then the unique weak solution u in $W^{1,m}(\Omega)$ of (2.2.10) satisfies

$$u \in L^{\frac{Nm}{\eta(N-m)}}(\Omega) \quad \text{and} \quad u|_{\partial\Omega} \in L^{\frac{(N-1)m}{\eta(N-m)}}(\partial\Omega).$$

We need an additional lemma, which essentially is a technical estimate.

Lemma 3.2.13. *Let $N \in \mathbb{N}$, $N \geq 2$, and let $m \in (1,N)$. For $\theta \in (0,1)$ we define $\eta_\theta \in (0,1)$ by*

$$\frac{1-\eta_\theta}{\eta_\theta} = (m-1)\frac{1-\theta}{\theta}. \tag{3.2.8}$$

Then

$$(2-m)(N-m) < m(1+\sqrt{m-1})^2 \tag{3.2.9}$$

if and only if

$$\eta_\theta - \theta < \frac{m}{N-m} \text{ for all } \theta \in (0,1). \tag{3.2.10}$$

Proof. If $m \geq 2$, then $\eta_\theta \leq \theta$ and both inequalities are trivially satisfied. Therefore we assume $m \in (1,2)$ in the proof.

Solving (3.2.8) for η_θ, we obtain that

$$\eta_\theta = \frac{\theta}{(m-1)(1-\theta)+\theta}.$$

We define a continuous function f on $[0,1]$ by

$$f(\theta) := \eta_\theta - \theta = \frac{\theta}{(m-1)(1-\theta)+\theta} - \theta = \frac{\theta}{(m-1)+(2-m)\theta} - \theta$$

and investigate in which situations we have

$$f(\theta) < \frac{m}{N-m} \text{ for all } \theta \in (0,1). \tag{3.2.11}$$

Since $f(0) = f(1) = 0$, by the necessary criterion for local extrema, condition (3.2.11) holds if and only if it holds for every $\theta_0 \in (0,1)$ satisfying

$$f'(\theta_0) = \frac{m-1}{((m-1)+(2-m)\theta_0)^2} - 1 = 0$$

i.e., for

$$\theta_0 = \frac{\sqrt{m-1}-(m-1)}{2-m} = \frac{\sqrt{m-1}}{1+\sqrt{m-1}} \in (0,1).$$

Now,

$$f(\theta_0) = \Big(\frac{1}{\sqrt{m-1}} - 1\Big)\theta_0 = \frac{1-\sqrt{m-1}}{1+\sqrt{m-1}} = \frac{2-m}{(1+\sqrt{m-1})^2}.$$

Thus,

$$f(\theta_0) < \frac{m}{N-m}$$

is equivalent to (3.2.9) and, as remarked before, also to (3.2.11), which merely is another way to write (3.2.10). □

For $m \in (1,2)$, the number η defined by (3.2.7) is larger than θ. Thus by applying Lemma 3.2.12 we do not gain as much regularity for the solutions as in Lemma 3.2.4.

Hence we need more restrictive growth assumptions on h than in Theorem 3.2.5. There is of course some freedom here about how fast we allow h to grow, and other growth conditions on h lead to conditions on m that are different from (3.2.12).

The choice (3.2.13) that we made here seems to be the most natural, in particular because it includes the standard example $h(u) = |u|^{m-2}u$.

Theorem 3.2.14. *Let $\Omega \subset \mathbb{R}^N$ be a bounded Lipschitz domain and $1 < m < N$. Assume that*

$$(2-m)(N-m) < m(1+\sqrt{m-1})^2. \tag{3.2.12}$$

Let h be a measurable function, and assume that there exist $\varepsilon > 0$, a constant $\mu \geq 0$, and ψ_9 in $L^{(N-1)/(m-1)+\varepsilon}(\partial\Omega)$ such that

$$|h(z,u)| \leq \mu|u|^{m-1} + \psi_9(z) \tag{3.2.13}$$

for all $z \in \partial\Omega$ and all $u \in \mathbb{R}$. Under Assumption 3.2.11, there exists $\alpha \in (0,1)$ such that every weak solution $u \in W^{1,m}(\Omega)$ of the Robin problem (1.9.6) is in $C^{0,\alpha}(\Omega)$ whenever f_0 is in $L^{N/m+\varepsilon}(\Omega)$, each f_j is in $L^{N/(m-1)+\varepsilon}(\Omega)$, and g is in $L^{(N-1)/(m-1)+\varepsilon}(\partial\Omega)$ for some $\varepsilon > 0$.

Proof. Let $u \in W^{1,m}(\Omega)$ be a weak solution of (1.9.6). We proceed exactly like in the proof of Theorem 3.2.5 and use the same notation. In particular, the set Θ is defined in the same way as before. There is only one part of the proof of Theorem 3.2.5 that does not carry over verbatim. In fact, we have to show that Θ is not of the form $(a,1]$ with $a \in [0,1)$.

To obtain a contradiction, assume that $\Theta = (a,1]$ with $a \in [0,1)$. Let $\theta \in (a,1]$ be arbitrary, and let ω be as in Assumption 3.2.1. Then the right hand side of (3.2.1) satisfies the assumptions of Lemma 3.2.12 for this θ. Thus

$$u \in L^{\frac{mN}{\eta_\theta(N-m)}}(\Omega) \quad \text{and} \quad u|_{\partial\Omega} \in L^{\frac{(N-1)m}{\eta_\theta(N-m)}}(\partial\Omega),$$

where η_θ is given by (3.2.7), and thus

$$|u|^{m-1} \in L^{\frac{mN}{\eta_\theta(N-m)(m-1)}}(\Omega) \quad \text{and} \quad |u|^{m-1}\Big|_{\partial\Omega} \in L^{\frac{(N-1)m}{\eta_\theta(N-m)(m-1)}}(\partial\Omega).$$

By Lemma 3.2.13,

$$\eta_a - a < \frac{m}{N-m},$$

hence

$$\eta_a(N-m)(m-1) < m(m-1) + a(N-m)(m-1) < m^2 + a(N-m)(m-1)$$

and thus

$$\frac{m(N-1)}{\eta_a(N-m)(m-1)} > \frac{m(N-1)}{m(m-1)+a(N-m)(m-1)} = q_a$$

and

$$\frac{mN}{\eta_a(N-m)(m-1)} > \frac{mN}{m^2+a(N-m)(m-1)} = p_a.$$

Hence, since η_θ depends continuously on θ, there exist $\theta > a$ and $\varepsilon > 0$ such that

$$\frac{m(N-1)}{\eta_\theta(N-m)(m-1)} \geq q_a + \varepsilon$$

and

$$\frac{mN}{\eta_\theta(N-m)(m-1)} \geq p_a + \varepsilon.$$

Thus $|u|^{m-1} \in L^{p_a+\varepsilon}(\Omega)$ and, by (3.2.13), $h(\cdot, u(\cdot)) \in L^{q_a+\varepsilon}(\Omega)$. We have shown that $a \in \Theta$, contradicting the assumption that $\Theta = (a,1]$.

Thus Θ is not of the form $(a, 1]$ with $a \in [0, 1)$. As already pointed out in the introductory remarks, we can now proceed like in the proof Theorem 3.2.5 to finish the proof. □

Remark 3.2.15. The condition (3.2.12) is rather strange and can probably be improved. However, it appears in a natural way because, by Lemma 3.2.13, it is equivalent to the fact that the bootstrapping works if h grows like (3.2.13). It is remarkable that (3.2.12) does not imply

$$m \geq \frac{2N}{N+2}, \tag{3.2.14}$$

which is a typical assumption for quasi-linear equations. In fact, for $N = 5$ condition (3.2.12) is equivalent to $m > \frac{5}{4}$.

Example 3.2.16 (m-Laplace equation)**.** Let $\Omega \subset \mathbb{R}^N$ be a bounded Lipschitz domain, let $m \in (1, 2)$, and assume that (3.2.12) holds. Let h be measurable and satisfy

$$|h(z, u)| \leq \mu |u|^{m-1} + \psi_9$$

with a constant $\mu \geq 0$ and a function ψ_9 in $L^{(N-1)/(m-1)+\varepsilon}(\partial\Omega)$ for some $\varepsilon > 0$, for example $h(z, u) = \beta |u|^{m-1}$ with $\beta \in L^\infty(\Omega)$. Let $f_0 \in L^{N/m+\varepsilon}(\Omega)$, $f_j \in L^{N/(m-1)+\varepsilon}(\Omega)$, and $g \in L^{(N-1)/(m-1)+\varepsilon}(\partial\Omega)$ for some $\varepsilon > 0$. Assume that $u \in W^{1,m}(\Omega)$ is a weak solution of

$$\begin{cases} -\operatorname{div}\Big(|\nabla u|^{m-2}\nabla u\Big) + \omega |u|^{m-2}u = f_0 - \sum\limits_{j=1}^{N} f_j, & \text{on } \Omega, \\ |\nabla u|^{m-2}\dfrac{\partial u}{\partial \nu} + h(z, u) = g + \sum\limits_{j=1}^{N} f_j \nu_j, & \text{on } \partial\Omega \end{cases}$$

for some $\omega \in \mathbb{R}$. Then u is Hölder continuous. Moreover, if h is continuous in its second argument and if ω is large enough, then for every right hand side there exists a weak solution. If in addition h satisfies (3.1.3), then every $\omega > 0$ is "large enough". Finally, if h is non-decreasing in its second argument, then the solution is unique.

Proof. We check Assumption 3.2.11 for $\beta = \frac{1}{m-1}$ with

$$A(x, u, p) = |p|^{m-2} p^T \text{ and } a(x, u, p) = 0.$$

The existence of solutions of the Neumann problem (2.2.10) is a special case of Example 2.2.10. The locally Lipschitz continuous dependence in the norm of $W^{1,m}(\Omega)$ of those solutions is the statement of Example 2.3.23. Finally, the Hölder regularity including the norm bound follows from Corollary 2.1.6 since the norm of the solution on $L^m(\Omega)$ occurring in (2.1.6) can be estimated using (2.2.11). Thus Assumption 3.2.11 is satisfied. Now Hölder regularity follows from Theorem 3.2.14. The additional claims can be deduced as in Example 3.2.9, compare also Example 2.2.10 to see that (2.2.13) holds with $\omega_0 = 0$. □

Remark 3.2.17. The bootstrapping argument is no longer needed if some other source provides bounds for the solutions. For example, if in the situation of Example 3.2.16, where we now allow arbitrary $m \in (1, \infty)$ (i.e., we no longer assume (3.2.12)), we have $h(z, u) = \beta |u|^{m-1}$ with a non-negative function $\beta \in L^\infty(\partial\Omega)$, then every weak solution $u \in W^{1,m}(\Omega)$ is bounded by a result due to Daners and Drábek [DD09], which is proved by Moser's iteration scheme. Hence $h(\cdot, u(\cdot))$ is bounded and we can apply Corollary 2.1.6 to (3.2.1) for $\omega := 0$, proving that the solution is Hölder continuous.

4 Parabolic equations

In this chapter we will combine the regularity result of Section 3.2 with semigroup theory to deduce that the corresponding parabolic problem is well-posed on $C(\overline{\Omega})$, i.e., that the part of the elliptic operator in $C(\overline{\Omega})$ generates a (linear or nonlinear) C_0-semigroup on $C(\overline{\Omega})$.

First we argue that the corresponding operator A generates a C_0-semigroup on $L^2(\Omega)$ which extrapolates to $L^\infty(\Omega)$. From this and the elliptic regularity result we obtain that the part A^c of A in $C(\overline{\Omega})$ generates a C_0-semigroup on the closure of the domain $D(A^c)$ of A^c. Then it only remains to show that A^c is densely defined in $C(\overline{\Omega})$. This is a consequence of the continuous dependence results in Section 3.2.

The described strategy works for linear and nonlinear problems with Robin boundary conditions, including Neumann boundary conditions as a special case, and also for Wentzell-Robin boundary conditions. In fact, for the latter the technical details are simpler than for the Robin problem. The linear case has already been treated by Warma [War06] in the case of the Laplace operator, and our transition from elliptic to parabolic results in Sections 4.1 and 4.2 is in the same spirit as his article. Here, however, we consider arbitrary second order, linear, elliptic operators and two model cases for nonlinear equations, which is made possible by our results in Chapter 3.

4.1 Linear equations with Robin boundary condition

Fix a bounded Lipschitz domain $\Omega \subset \mathbb{R}^N$ and coefficients a_{ij}, b_j, c_i, and d in $L^\infty(\Omega)$ such that there exists $\nu > 0$ with the property that

$$\sum_{i,j=1}^{N} a_{ij}\xi_i\xi_j \geq \nu|\xi|^2$$

for all $\xi \in \mathbb{R}^N$. To simplify notation we introduce the shorthands

$$Lu := -\sum_{j=1}^{N} D_j\Big(\sum_{i=1}^{N} a_{ij}D_iu + b_ju\Big) + \Big(\sum_{i=1}^{N} c_iD_iu + du\Big)$$

and

$$\frac{\partial u}{\partial \nu_L} := \sum_{j=1}^{N}\Big(\sum_{i=1}^{N} a_{ij}D_iu + b_ju\Big)\nu_j,$$

where as usual $\nu = (\nu_j)$ denotes the outer unit normal to Ω on $\partial\Omega$. We do not distinguish between a continuous function u from $[0,\infty)$ to $L^2(\Omega)$ and the associated real-valued function $(t,x) \mapsto u(t)(x)$ on $[0,\infty) \times \Omega$. Let β be in $L^\infty(\partial\Omega)$, and let u_0 be in $\mathrm{C}(\overline{\Omega})$. The goal of this section is to prove that the linear parabolic Robin problem

$$\begin{cases} \dot{u}(t,x) = -Lu(t,x), & t > 0,\ x \in \Omega, \\ \frac{\partial u}{\partial \nu_L}(t,z) + \beta u(t,z) = 0, & t > 0,\ z \in \partial\Omega, \\ u(0,x) = u_0(x), & x \in \Omega, \end{cases} \tag{4.1.1}$$

is well-posed in the space $\mathrm{C}(\overline{\Omega})$. More precisely, we will show the unique mild solution u in the sense of $L^2(\Omega)$ is continuous as a function from $[0,\infty)$ to $\mathrm{C}(\overline{\Omega})$, or, which is the same, is jointly continuous in its time and space variables.

Let H denote the Hilbert space $L^2(\Omega)$. For the dense subspace $V := W^{1,2}(\Omega)$, define the form $\mathfrak{a}$ on $V \times V$ by

$$\mathfrak{a}(u,v) := \sum_{i,j=1}^N \int_\Omega a_{ij} D_i u\, D_j v + \sum_{j=1}^N \int_\Omega b_j u\, D_j v + \sum_{i=1}^N \int_\Omega c_i D_i u\, v + \int_\Omega duv + \int_{\partial\Omega} \beta uv.$$

As in the proof of Proposition 3.1.3, we obtain that $\mathfrak{a}$ is continuous and elliptic. Thus, by Remark 1.7.8, it is associated with a closed linear operator A on $L^2(\Omega)$ such that $-A$ generates a C_0-semigroup $(T(t))_{t\geq 0}$ on $L^2(\Omega)$.

The following remark shows that A is in fact a realization of L in $L^2(\Omega)$ with Robin boundary conditions, thus justifying Definition 4.1.2.

Remark 4.1.1. Let all functions be smooth enough to apply the divergence theorem. For simplicity, we set $b_j = c_i = d = 0$. Now if $u \in D(A)$ and $v \in W^{1,2}(\Omega)$, then

$$\begin{aligned} \int_\Omega Au\, v = (Au \mid v)_{L^2(\Omega)} = \mathfrak{a}(u,v) &= \sum_{i,j=1}^N \int_\Omega a_{ij} D_i u D_j v + \int_{\partial\Omega} \beta uv \\ &= \int_{\partial\Omega} \Big(\sum_{i,j=1}^N a_{ij} D_i u \nu_j + \beta u \Big) v - \sum_{j=1}^N \int_\Omega D_j \Big(\sum_{i=1}^N a_{ij} D_i u \Big) v \\ &= \int_{\partial\Omega} \Big(\frac{\partial u}{\partial \nu_L} + \beta u \Big) v + \int_\Omega Lu\, v. \end{aligned}$$

Thus this identity holds in particular for all test functions $v \in \mathrm{C}_c^\infty(\Omega)$, for which the boundary integral vanishes. Since the test functions are dense in $L^2(\Omega)$, we deduce that $Au = Lu$. Thus the boundary integral equals zero for all $v \in W^{1,2}(\Omega)$. Since the traces of functions in $W^{1,2}(\Omega)$ are dense in $L^2(\partial\Omega)$, this implies $\frac{\partial u}{\partial \nu_L} + \beta u = 0$ i.e., u satisfies Robin boundary conditions.

Definition 4.1.2. We say that a continuous function u from $[0,\infty)$ to $L^2(\Omega)$ is a *mild solution of* (4.1.1) for some $u_0 \in L^2(\Omega)$ if u is a mild solution of the abstract Cauchy problem

$$\begin{cases} \dot{u}(t) = -Au(t), & t > 0, \\ u(0) = u_0 \end{cases} \tag{4.1.2}$$

in the sense of Definition 1.7.4, i.e., $u(t) = T(t)u_0$ for the semigroup $(T(t))_{t\geq 0}$ that is generated by $-A$, see Theorem 1.7.5. We call A the *realization of L in $L^2(\Omega)$ with Robin boundary conditions* and $(T(t))_{t\geq 0}$ the corresponding *Robin semigroup*.

It follows from Theorem 1.7.5 that for every $u_0 \in L^2(\Omega)$ problem (4.1.1) has a unique mild solution. Now if $u\colon [0,\infty) \to H$ is a mild solution of (4.1.2), then u is differentiable on $(0,\infty)$, $u(t) \in D(A)$ for all $t > 0$, and $\dot{u}(t) = -Au(t)$ for all $t > 0$ by Theorem 1.7.7. Thus $\dot{u}(t) = -Lu(t)$ on Ω and $\frac{\partial u(t)}{\partial \nu_L} + \beta u(t) = 0$ on $\partial\Omega$ for all $t > 0$, i.e., u satisfies (4.1.1) in a weak sense.

It is in general non-trivial to show that the semigroup T leaves $L^\infty(\Omega)$ invariant, even though in some special cases this follows from Corollary 1.7.10. For example, Corollary 1.7.10 applies if $b_j \in W_0^{1,\infty}(\Omega)$ and $\beta \geq 0$, and this fact has been used by Arendt and ter Elst [AtE97] to prove Gaussian estimates in this case. Unfortunately, in general the equivalent conditions of Corollary 1.7.10 are not satisfied for $(T(t))_{t\geq 0}$ even after exponential rescaling as Example 4.1.3 shows. Anyway, by different methods Daners [Dan00a] managed to obtain exponential bounds in $L^\infty(\Omega)$, and we cite his results in Lemma 4.1.4.

Example 4.1.3. Let $N = 1$, $\Omega = (-1,1)$, $a_{11} = 1$, $b_1(x) = \operatorname{sgn}(x)$, $c_1 = 0$, $d = 0$, and $\beta = 0$. Then there exists no $\omega \in \mathbb{R}$ such that $(\mathrm{e}^{-\omega t}\, T(t))_{t\geq 0}$ is contractive in $L^\infty(\Omega)$.

Proof. Assume to the contrary that $(\mathrm{e}^{-\omega t}\, T(t))_{t\geq 0}$ is contractive in the norm of $L^\infty(\Omega)$ for some $\omega \in \mathbb{R}$. Without loss of generality we can assume $\omega \geq 0$. Since this semigroup is associated with the form $(u,v) \mapsto \mathfrak{a}(u,v) + \omega\,(u \mid v)_{L^2(\Omega)}$, Corollary 1.7.10 implies that

$$\mathfrak{a}\Big(u \wedge 1, (u-1)^+\Big) + \omega\, \Big(u \wedge 1 \mid (u-1)^+\Big)_{L^2(\Omega)} \geq 0 \tag{4.1.3}$$

for all positive $u \in W^{1,2}(-1,1)$. Using Lemma 1.2.6 to calculate the weak derivatives of $u \wedge 1$ and $(u-1)^+$, we see that $(u \wedge 1)' = u'$ and $((u-1)^+)' = 0$ almost everywhere on $\{u \leq 1\}$, and that $(u \wedge 1)' = 0$ and $((u-1)^+)' = u'$ almost everywhere on $\{u \geq 1\}$. Hence (4.1.3) reads as

$$-\int_{[-1,0]\cap\{u\geq 1\}} u' + \int_{[0,1]\cap\{u\geq 1\}} u' + \omega \int_{[-1,1]\cap\{u\geq 1\}} (u-1) \geq 0$$

for all positive $u \in W^{1,2}(-1,1)$. If we set in particular $u_n(x) := 2(1-x^2)^n$ and pick α_n such that $u_n(\pm\alpha_n) = 1$, i.e., $\alpha_n := (1 - 2^{-1/n})^{1/2}$, then

$$0 \leq -\int_{-\alpha_n}^{0} u_n' + \int_0^{\alpha_n} u_n' + \omega \int_{-\alpha_n}^{\alpha_n} (u_n - 1) \leq -2 + \omega \int_{-\alpha_n}^{\alpha_n} 1 = -2 + 2\omega\alpha_n$$

for all $n \in \mathbb{N}$. However, since $\alpha_n \to 0$ as $n \to \infty$, this cannot hold for large n, a contradiction. Thus $(\mathrm{e}^{-\omega t}\, T(t))_{t\geq 0}$ is not contractive in $L^\infty(\Omega)$. □

Lemma 4.1.4. *There exist $M \geq 1$ and $\omega \in \mathbb{R}$ such that*

$$\|T(t)u\|_{L^\infty(\Omega)} \leq M\,\mathrm{e}^{\omega t}\,\|u\|_{L^\infty(\Omega)}$$

for all $t \geq 0$ and all $u \in L^\infty(\Omega)$.

Proof. For $\beta \geq 0$, Daners [Dan00a, Proposition 7.1] has proved precisely this statement by obtaining Gaussian upper bounds for the associated kernels. By rewriting the equation in an appropriate way, the same author [Dan09] showed that the general result then follows for arbitrary bounded β. □

Lemma 4.1.5. *The part $A^c := A|_{\mathrm{C}(\overline{\Omega})}$ of A in $\mathrm{C}(\overline{\Omega})$, i.e., the restriction of A to*

$$D(A^c) := \left\{u \in D(A) \cap \mathrm{C}(\overline{\Omega}) : Au \in \mathrm{C}(\overline{\Omega})\right\},$$

considered as an operator on $\mathrm{C}(\overline{\Omega})$, is closed. Moreover, if $\lambda + A$ is boundedly invertible for some $\lambda \in \mathbb{R}$, then also $\lambda + A^c$ is boundedly invertible, i.e., $\varrho(A) \subset \varrho(A^c)$. In this case, $\mathrm{C}(\overline{\Omega})$ is invariant under $R(\lambda, -A)$, and $R(\lambda, -A^c) = R(\lambda, -A)|_{\mathrm{C}(\overline{\Omega})}$. Finally, if λ is sufficiently large, then $\lambda + A^c$ is invertible and

$$\|R(\lambda, -A^c)^{(n)}\| \leq \frac{Mn!}{(\lambda - \omega)^{n+1}}$$

for all $n \in \mathbb{N}$, where ω is as in Lemma 4.1.4 and where $R(\lambda, -A^c)^{(n)}$ denotes the nth derivative of the analytic function $\lambda \mapsto R(\lambda, -A^c)$.

Proof. If (u_n) is a sequence that converges in $\mathrm{C}(\overline{\Omega})$ to u such that $(A^c u_n)$ converges in $\mathrm{C}(\overline{\Omega})$ to v, then the two sequences converge also in $L^2(\Omega)$. Since A, being the negative of a generator of a C_0-semigroup, is closed, this implies that $u \in D(A)$ and $Au = v$. But since u and v are in $\mathrm{C}(\overline{\Omega})$ by assumption, this implies $u \in D(A^c)$. Thus A^c is closed.

Let $\lambda + A$ be invertible. Then $\lambda + A^c$ is injective. Moreover, for given $f \in \mathrm{C}(\overline{\Omega})$, there exists $u \in D(A)$ such that $(\lambda + A)u = f$. Then u is a weak solution in $W^{1,2}(\Omega)$ of

$$\begin{cases} Lu + \lambda u = f & \text{on } \Omega, \\ \dfrac{\partial u}{\partial \nu_L} + \beta u = 0 & \text{on } \partial\Omega. \end{cases}$$

Hence u is continuous by Example 3.2.7, showing that $u \in D(A^c)$. Thus $\lambda + A^c$ is surjective and consequently invertible by the closed graph theorem.

By Definition 1.7.1, there exists ω_0 such that $\lambda + A$ is invertible for $\lambda > \omega_0$, and by differentiating (1.7.1) with respect to λ,

$$R(\lambda, -A)^{(n)}u = (-1)^n \int_0^\infty s^n\,\mathrm{e}^{-\lambda s}\,T(s)u\,\mathrm{d}s$$

for all $u \in L^2(\Omega)$ and $\lambda > \omega_0$. Thus from Lemma 4.1.4 we obtain by repeated integration by parts that

$$\|R(\lambda, -A)^{(n)}u\|_{L^\infty(\Omega)} \leq M \int_0^\infty s^n \, \mathrm{e}^{(\omega-\lambda)s} \, \mathrm{d}s \; \|u\|_{L^\infty(\Omega)} = \frac{Mn!}{(\lambda-\omega)^{n+1}} \|u\|_{L^\infty(\Omega)}$$

for all $u \in L^\infty(\Omega)$. This proves the claim. □

Lemma 4.1.6. *The operator A^c is densely defined on $\mathrm{C}(\overline{\Omega})$.*

Proof. Since $\mathrm{C}^\infty(\overline{\Omega})$ is dense in $\mathrm{C}(\overline{\Omega})$ by the Stone-Weierstrass theorem, it suffices to show that every $u \in \mathrm{C}^\infty(\overline{\Omega})$ is in the closure of $D(A^c)$. Thus let $u \in \mathrm{C}^\infty(\overline{\Omega})$ be arbitrary and fix $\varepsilon > 0$.

Pick $q > N$, let ω_0 be as in Proposition 3.1.3, and fix $\omega > \omega_0$. Let h_d be as in Lemma 1.3.12, i.e., h_d is a smooth vector field such that $h_d \cdot \nu \geq 1$ almost everywhere on $\partial\Omega$. Since the traces of smooth functions are dense in $L^{q-1}(\partial\Omega)$ by the Stone-Weierstrass theorem, there exists $h \in \mathrm{C}^\infty(\mathbb{R}^N; \mathbb{R}^N)$ such that

$$\left\| h - \tfrac{\beta u h_d}{h_d \cdot \nu} \right\|_{L^{q-1}(\partial\Omega;\mathbb{R}^N)} < \varepsilon.$$

Hence $g := h \cdot \nu - \beta u$ satisfies $\|g\|_{L^{q-1}(\partial\Omega)} < \varepsilon$. Since $\mathrm{C}_c^\infty(\Omega)$ is dense in $L^p(\Omega)$, there exist $k_0, k_1, \dots, k_N$ in $\mathrm{C}_c^\infty(\Omega)$ such that the functions

$$f_0 := k_0 - \sum_{i=1}^N c_i D_i u - du - \omega u \quad \text{and} \quad f_j := -h_j - k_j - \sum_{i=1}^N a_{ij} D_i u - b_j u$$

satisfy $\|f_0\|_{L^{q/2}(\Omega)} < \varepsilon$ and $\|f_j\|_{L^q(\Omega)} < \varepsilon$. Define $\psi \in \mathrm{C}^\infty(\overline{\Omega})$ by

$$\psi := k_0 + \sum_{j=1}^N D_j k_j + \operatorname{div}(h)$$

and let $v := R(\omega, -A)$, i.e., v is the unique solution of

$$\begin{cases} Lv + \omega v = \psi & \text{on } \Omega, \\ \dfrac{\partial v}{\partial \nu_L} + \beta v = 0 & \text{on } \partial\Omega \end{cases}$$

which exists by Proposition 3.1.3. By the divergence theorem,

$$\int_\Omega \psi w = \int_\Omega k_0 w + \int_{\partial\Omega} (h \cdot \nu) w - \sum_{j=1}^N \int_\Omega (h_j + k_j) D_j w$$

for every $w \in \mathrm{C}^\infty(\overline{\Omega})$. Hence, by definition of v and $\mathfrak{a}$, we have

$$\mathfrak{a}(v-u, w) + \omega\,(v - u \mid w)_{L^2(\Omega)}$$

$$= \int_\Omega \psi w - \mathfrak{a}(u,w) - \omega\,(u \mid w)_{L^2(\Omega)} = \int_\Omega f_0 w + \sum_{j=1}^N \int_\Omega f_j D_j w + \int_{\partial\Omega} g w$$

for all $w \in C^\infty(\overline{\Omega})$, i.e., we have in a weak sense that

$$\begin{cases} L(v-u) + \omega(v-u) = f_0 - \sum_{j=1}^N D_j f_j & \text{on } \Omega, \\ \dfrac{\partial(v-u)}{\partial\nu_L} + \beta(v-u) = g + \sum_{j=1}^N f_j \nu_j & \text{on } \partial\Omega. \end{cases} \tag{4.1.4}$$

Since by Example 3.2.7 the solution operator maps appropriate right-hand-side spaces into $C^{0,\alpha}(\Omega)$, $\alpha \in (0,1)$, and thus in particular into $C(\overline{\Omega})$, the closed graph theorem asserts that the solution operator is bounded between these spaces. Since the right hand side of (4.1.4) is small by construction, we can conclude that

$$\|u - v\|_{L^\infty(\Omega)} \le c\varepsilon$$

for a constant c that does not depend on ε, u, or v. Since v is in $D(A+\omega) = D(A)$ by definition, letting ε tend to zero we find a family of functions in $D(A)$ converging to u in the norm of $C(\overline{\Omega})$, which proves the claim. □

Theorem 4.1.7. *Problem 4.1.1 is well-posed on $C(\overline{\Omega})$, meaning that the operator $-A^c$ defined in Lemma 4.1.5 generates a C_0-semigroup on $C(\overline{\Omega})$. Thus for every $u_0 \in C(\overline{\Omega})$ the unique mild solution $u\colon [0,\infty) \to L^2(\Omega)$ of (4.1.1) is continuous as a function from $[0,\infty)$ to $C(\overline{\Omega})$. In other words, we can choose representatives of $u(t)$, $t \ge 0$, in such a way that $(t,x) \mapsto u(t)(x)$ is continuous on $[0,\infty) \times \overline{\Omega}$.*

Proof. Since by Lemmata 4.1.5 and 4.1.6 the operator $-A^c$ satisfies the assumptions of Theorem 1.7.2, this operator generates a C_0-semigroup $(T^c(t))_{t\ge 0}$ on $C(\overline{\Omega})$. Since semigroups can be expressed in terms of their generators via (1.7.3) and since the resolvents of A^c and A coincide on $C(\overline{\Omega})$, $T^c(t)$ is the restriction of $T(t)$ to $C(\overline{\Omega})$. Thus the mild solution $u(t) = T(t)u_0 = T^c(t)u_0$ of (4.1.1) is continuous with values in $C(\overline{\Omega})$. □

4.2 Linear equations with Wentzell-Robin boundary conditions

Let $\Omega \subset \mathbb{R}^N$ be a bounded Lipschitz domain, let L be as in the previous section, let β be a function in $L^\infty(\partial\Omega)$, let $\gamma \in L^\infty(\partial\Omega)$ be bounded from below by $\gamma_0 > 0$, and let u_0 be in $C(\overline{\Omega})$. The goal of this section is to prove that the linear parabolic Wentzell-Robin problem

$$\begin{cases} \dot{u}(t,x) = -Lu(t,x) & t>0, x\in\Omega, \\ -Lu(t,z)+\gamma\frac{\partial u(t,z)}{\partial\nu_L}+\beta u(t,z)=0, & t>0, z\in\partial\Omega, \\ u(0,x)=u_0(x), & x\in\Omega \end{cases} \tag{4.2.1}$$

is well-posed in the space $\mathrm{C}(\overline{\Omega})$. More precisely, we will show the unique mild solution u in the sense of $L^2(\Omega)$ is continuous as a function from $[0,\infty)$ to $\mathrm{C}(\overline{\Omega})$, or, which is the same, the solution is jointly continuous in its time and space variables.

To study (4.2.1) in a Hilbert space setting, we use the form method approach proposed by Arendt, Metafune, Pallara, and Romanelli [AMPR03]. There are other ways to attack the problem, but for our purposes the form method turns out to be very convenient since the arguments of Section 4.1 carry over to a large extent. The appropriate space for (4.2.1) is the Hilbert space

$$\mathcal{H} := L^2(\Omega)\oplus_2 L^2(\partial\Omega,\tfrac{\mathrm{d}\sigma}{\gamma})$$
$$(u\mid v)_{\mathcal{H}} := \int_\Omega uv + \int_{\partial\Omega}\frac{uv}{\gamma},$$

where by the assumption on γ the norm of $L^2(\partial\Omega,\frac{\mathrm{d}\sigma}{\gamma})$ is equivalent to the usual norm of $L^2(\partial\Omega)=L^2(\partial\Omega,\mathrm{d}\sigma)$ stemming from the surface measure σ on $\partial\Omega$. For u and v in the subspace

$$\mathcal{V} := \left\{(u,u|_{\partial\Omega}) : u\in W^{1,2}(\Omega)\right\}\cong W^{1,2}(\Omega)$$

of $\mathcal{H}$, which we equip with the norm $\|(u,u|_{\partial\Omega})\|_{\mathcal{V}} := \|u\|_{W^{1,2}(\Omega)}$, we define

$$\mathfrak{a}\Big((u,u|_{\partial\Omega}),(v,v|_{\partial\Omega})\Big) := \sum_{i,j=1}^N\int_\Omega a_{ij}D_iu\,D_jv+\sum_{j=1}^N\int_\Omega b_ju\,D_jv+\sum_{i=1}^N\int_\Omega c_iD_iu\,v$$
$$+\int_\Omega duv+\int_{\partial\Omega}\frac{\beta}{\gamma}uv.$$

Since $\|(u,u|_{\partial\Omega})\|_{\mathcal{H}}\geq\|u\|_{L^2(\Omega)}$, the same estimates as in the proof of Proposition 3.1.3 show that $\mathfrak{a}$ is continuous and coercive. Moreover, since the traces of smooth functions, i.e., functions in $\mathrm{C}^\infty(\overline{\Omega})$, are dense in $L^2(\partial\Omega)$ by the Stone-Weierstrass theorem, and since $\mathrm{C}_c^\infty(\Omega)$ is dense in $L^2(\Omega)$, the form domain $\mathcal{V}$ is densely and continuously embedded into $\mathcal{H}$. Thus, by Remark 1.7.8, the form $\mathfrak{a}$ is associated with a closed linear operator A such that $-A$ generates a C_0-semigroup $(T(t))_{t\geq0}$ on $\mathcal{H}$.

The following remark shows that A is in fact a realization of L in $L^2(\Omega)$ with Wentzell-Robin-like boundary conditions. This is used to justify Definition 4.2.2.

Remark 4.2.1. Let all functions be smooth enough to apply the divergence theorem. For simplicity, we set $b_j=c_i=d=0$. Now if $(u,u|_{\partial\Omega})\in D(A)\subset\mathcal{V}$ with $(f,g):=Au\in\mathcal{H}$

and $(v, v|_{\partial\Omega}) \in \mathcal{V}$, then

$$\begin{aligned}\int_\Omega fv + \int_{\partial\Omega} \frac{gv}{\gamma} &= (Au \mid (v, v|_{\partial\Omega}))_{\mathcal{H}} = \mathfrak{a}\big((u, u|_{\partial\Omega}), (v, v|_{\partial\Omega})\big)\\ &= \sum_{i,j=1}^N \int_\Omega a_{ij} D_i u\, D_j v + \int_{\partial\Omega} \frac{\beta}{\gamma} uv\\ &= \int_{\partial\Omega} \Big(\sum_{i,j=1}^N a_{ij} D_i u\, \nu_j + \frac{\beta}{\gamma} u\Big) v - \sum_{j=1}^N \int_\Omega D_j \Big(\sum_{i=1}^N a_{ij} D_i u\Big) v\\ &= \int_{\partial\Omega} \Big(\frac{\partial u}{\partial \nu_L} + \frac{\beta}{\gamma} u\Big) v + \int_\Omega Lu\, v.\end{aligned}$$

Since this holds in particular for $v \in C_c^\infty(\Omega)$, for which the boundary terms disappear, and since such functions are dense in $L^2(\Omega)$, we deduce that $f = Lu$. Since now the interior integrals cancel and since the traces of functions in $W^{1,2}(\Omega)$ are dense in $L^2(\partial\Omega)$, this implies that $g = \gamma\frac{\partial u}{\partial \nu_L} + \beta u$. Thus $Au = (Lu, \gamma\frac{\partial u}{\partial \nu_L} + \beta u)$.

Definition 4.2.2. We say that a continuous function u from $[0, \infty)$ to $\mathcal{H}$ is a *mild solution of* (4.2.1) for some $u_0 \in \mathcal{H}$ if it is a mild solution of the abstract Cauchy problem

$$\begin{cases} \dot{u} = -Au(t), & t > 0,\\ u(0) = u_0 \end{cases} \tag{4.2.2}$$

in the sense of Definition 1.7.4, i.e., $u(t) = T(t)u_0$ for the semigroup $(T(t))_{t\geq 0}$ generated by $-A$, see Theorem 1.7.5. We call A the *realization of L in $L^2(\Omega)$ with Wentzell-Robin boundary conditions* and $(T(t))_{t\geq 0}$ the corresponding *Wentzell-Robin semigroup*.

It follows from Theorem 1.7.5 that for every $u_0 \in \mathcal{H}$ problem (4.2.1) has a unique mild solution. Now if $U\colon [0, \infty) \to \mathcal{H}$ is a mild solution of (4.2.2), then U is differentiable on $(0, \infty)$, $U(t) \in D(A) \subset \mathcal{V}$, and $\dot{U}(t) = -AU(t)$ by Theorem 1.7.7. Thus

$$U(t) = (u(t), u(t)|_{\partial\Omega})$$

for every $t > 0$ with some $u(t) \in W^{1,2}(\Omega)$, and

$$(\dot{u}(t), \dot{u}(t)|_{\partial\Omega}) = \dot{U}(t) = -AU(t) = (-Lu(t), -\gamma\frac{\partial u(t)}{\partial \nu_L} - \beta u(t))$$

for $t > 0$, i.e., $\dot{u}(t) = -Lu(t)$ and

$$Lu(t)|_{\partial\Omega} = -\dot{u}(t)|_{\partial\Omega} = \gamma\frac{\partial u(t)}{\partial \nu_L} + \beta u(t).$$

This means that u satisfies (4.2.1) in a weak sense.

The following lemmata contain the only real differences to the Robin case. Again, it is in general non-trivial to show that the semigroup leaves the space of bounded functions invariant. This time, however, Daners' strategy seems not to work. Thus we restrict ourselves to a special case, namely we assume the coefficients b_j to be Lipschitz continuous. Even though this might not be a harsh restriction, it is somewhat unsatisfactory to introduce such a strange condition. It is of course motivated by the fact that under this hypothesis the Wentzell-Robin semigroup is contractive on the space of bounded functions, which is not true in general as we can see by taking the same coefficients as in Example 4.1.3.

Lemma 4.2.3. *Assume that $b_j \in W^{1,\infty}(\Omega)$, $j = 1, \dots, N$. Then there exists $\omega \in \mathbb{R}$ such that*

$$\|T(t)U\|_{L^\infty(\Omega)\oplus_\infty L^\infty(\partial\Omega)} \le e^{\omega t} \|U\|_{L^\infty(\Omega)\oplus_\infty L^\infty(\partial\Omega)}$$

for all $t \ge 0$ and all U in $L^\infty(\Omega) \oplus_\infty L^\infty(\partial\Omega) \subset \mathcal{H}$, where this space is equipped with the norm

$$\|(u,v)\|_{L^\infty(\Omega)\oplus_\infty L^\infty(\partial\Omega)} := \max\Big\{\|u\|_{L^\infty(\Omega)}, \|v\|_{L^\infty(\partial\Omega)}\Big\}.$$

Proof. If we equip $\overline{\Omega}$ with the measure

$$\mu(A) := \int_\Omega \mathbb{1}_{A\cap\Omega} + \int_{\partial\Omega} \frac{\mathbb{1}_{A\cap\Omega}}{\gamma},$$

then $L^2(\overline{\Omega},\mu) = \mathcal{H}$ and $L^\infty(\overline{\Omega},\mu) = L^\infty(\Omega) \oplus_\infty L^\infty(\partial\Omega)$.

By continuity of the trace operator, the operations $u \mapsto |u|$, $u \mapsto u^+$, $u \mapsto u \wedge 1$, and $u \mapsto \operatorname{sgn}(u)$ are respected by the trace operator for every $u \in W^{1,2}(\Omega)$ since this is true for continuous functions. Thus, by Lemma 1.2.6, we obtain that for $U = (u, u|_{\partial\Omega}) \in \mathcal{V}$ we have

$$(|U| \wedge 1)\operatorname{sgn}(U) = \Big((|u| \wedge 1)\operatorname{sgn}(u), (|u|\big|_{\partial\Omega} \wedge 1)\operatorname{sgn}(u|_{\partial\Omega})\Big) \in \mathcal{V}.$$

Since

$$D_i\Big((|u| \wedge 1)\operatorname{sgn}(u)\Big) = D_i u\, \mathbb{1}_{\{|u|<1\}} \quad\text{and}\quad D_j\Big((|u|-1)^+\operatorname{sgn}(u)\Big) = D_j u\, \mathbb{1}_{\{|u|>1\}},$$

by (1.2.1), we obtain that

$$\begin{aligned}&\mathfrak{a}\Big((|U| \wedge 1)\operatorname{sgn}(U), (|U|-1)^+\operatorname{sgn}(U)\Big)\\ &\quad= \sum_{j=1}^N \int_\Omega b_j\, D_j\Big((|u|-1)^+\operatorname{sgn}(u)\Big) + \int_\Omega d(|u|-1)^+ + \int_{\partial\Omega} \frac{\beta}{\gamma}(|u|-1)^+\end{aligned}$$

for $U = (u, u|_{\partial\Omega}) \in \mathcal{V}$. Now, by the divergence theorem,

$$\sum_{j=1}^N \int_\Omega b_j D_j\Big((|u|-1)^+\operatorname{sgn}(u)\Big)$$

$$= \sum_{j=1}^{N} \int_{\partial\Omega} b_j \nu_j \, (|u|-1)^+ \operatorname{sgn}(u) - \sum_{j=1}^{N} \int_{\Omega} D_j b_j \, (|u|-1)^+ \operatorname{sgn}(u).$$

From these two identities and the fact that the coefficients and the functions $D_j b_j$ are bounded, we obtain that there exists $\omega \geq 0$ such that

$$\begin{aligned} \Big|\mathfrak{a}\big((|U| \wedge 1)\operatorname{sgn}(U), (|U|-1)^+ \operatorname{sgn}(U)\big)\Big| &\leq \omega \int_{\Omega} (|u|-1)^+ + \omega \int_{\partial\Omega} (|u|-1)^+ \\ &= \omega \left((|U| \wedge 1)\operatorname{sgn}(U) \mid (|U|-1)^+ \operatorname{sgn}(U)\right)_{\mathcal{H}} \end{aligned}$$

for all $U = (u, u|_{\partial\Omega}) \in \mathcal{V}$. Thus the form $\mathfrak{a}^{\omega}$ defined on $\mathcal{V} \times \mathcal{V}$ by

$$\mathfrak{a}^{\omega}(U,V) := \mathfrak{a}(U,V) + \omega\,(U \mid V)_{\mathcal{H}}$$

satisfies (1.7.5) for the measure space $(\overline{\Omega}, \mu)$. By Corollary 1.7.10, since $\mathfrak{a}^{\omega}$ is associated with the semigroup $(\mathrm{e}^{-\omega t} T(t))_{t \geq 0}$, this proves the claim. $\square$

If we want to prove results about the Wentzell-Robin semigroup on the space of continuous functions, we have to embed $\mathrm{C}(\overline{\Omega})$ into $\mathcal{H}$. The natural way to do this is to consider the isomorphic copy

$$\begin{aligned} \mathscr{C} &:= \left\{(u, u|_{\partial\Omega}) : u \in \mathrm{C}(\overline{\Omega})\right\} \\ \|(u, u|_{\partial\Omega})\|_{\mathscr{C}} &:= \|u\|_{L^{\infty}(\Omega)} \end{aligned}$$

of $\mathrm{C}(\overline{\Omega})$ in $\mathcal{H}$. The space $\mathscr{C}$ is a closed subspace of $L^{\infty}(\Omega) \oplus_{\infty} L^{\infty}(\partial\Omega)$ and carries the induced norm.

Lemma 4.2.4. *The part A^c of A in $\mathscr{C}$ is closed. Moreover, if $\lambda + A$ is boundedly invertible for some $\lambda \in \mathbb{R}$, then also $\lambda + A^c$ is boundedly invertible, i.e., $\varrho(A) \subset \varrho(A^c)$. In this case, $\mathscr{C}$ is invariant under $R(\lambda, -A)$, and $R(\lambda, -A^c) = R(\lambda, -A)|_{\mathscr{C}}$. If b_j is in $W^{1,\infty}(\Omega)$ for all $j = 1, \dots, N$ and if λ is sufficiently large, then $\lambda + A^c$ is invertible and*

$$\|R(\lambda, -A^c)^{(n)}\| \leq \frac{n!}{(\lambda - \omega)^{n+1}}$$

for all $n \in \mathbb{N}$, where ω is as in Lemma 4.2.3.

Proof. The closedness follows as in Lemma 4.1.5.

Let $\lambda + A$ be invertible. Then $\lambda + A^c$ is injective, too. Moreover, for given $(f, f|_{\partial\Omega}) \in \mathscr{C}$, there exists $(u, u|_{\partial\Omega})$ in $D(A)$ such that $(\lambda + A)(u, u|_{\partial\Omega}) = (f, f|_{\partial\Omega})$. Then u is a weak solution in $W^{1,2}(\Omega)$ of

$$\begin{cases} Lu + \lambda u = f & \text{on } \Omega, \\ \dfrac{\partial u}{\partial \nu_L} + \dfrac{\beta}{\gamma} u + \dfrac{\lambda}{\gamma} u = \dfrac{f}{\gamma} & \text{on } \partial\Omega. \end{cases}$$

Thus u is continuous by Example 3.2.7, showing that $(u, u_{\partial\Omega}) \in D(A^c)$. Hence $\lambda + A^c$ is surjective and consequently invertible by the closed graph theorem.

The rest of the claim follows from Lemma 4.2.3 like in Lemma 4.1.5 it followed from Lemma 4.1.4. □

Lemma 4.2.5. *The operator A^c is densely defined in $\mathscr{C}$.*

Proof. The space $\mathscr{C}^\infty := \left\{(u, u|_{\partial\Omega}) : u \in \mathrm{C}^\infty(\overline{\Omega})\right\}$ is dense in $\mathscr{C}$ by the Stone-Weierstrass theorem. Let $\varepsilon > 0$, pick $q > N$, let ω_0 be as in Proposition 3.1.3, and fix $\omega > \omega_0$. Since $\mathscr{C}^\infty$ is dense in $\mathscr{C}$, there exists h_0 in $\mathrm{C}^\infty(\overline{\Omega})$ such that

$$g := \frac{h_0|_{\partial\Omega} - \beta u - \omega u}{\gamma}$$

satisfies $\|g\|_{L^{q-1}(\partial\Omega)} < \varepsilon$. Since $\mathrm{C}_c^\infty(\Omega)$ is dense in $L^q(\Omega)$, there exist $k_0, k_1, \dots, k_N$ in $\mathrm{C}_c^\infty(\Omega)$ such that

$$f_0 := k_0 + h_0 - \sum_{i=1}^N c_i D_i u - du - \omega u \quad \text{and} \quad f_j := -k_j - \sum_{i=1}^N a_{ij} D_i u - b_j u$$

satisfy $\|f_0\|_{L^{q/2}(\Omega)} < \varepsilon$ and $\|f_j\|_{L^q(\Omega)} < \varepsilon$. Now define

$$\psi := k_0 + \sum_{j=1}^N D_j k_j + h_0.$$

Let $v := -R(-\omega, A)(\psi, \psi|_{\partial\Omega})$, i.e., v is the unique weak solution of

$$\begin{cases} Lv + \omega v = \psi & \text{on } \Omega, \\ \dfrac{\partial v}{\partial \nu_L} + \dfrac{\beta}{\gamma} v + \dfrac{\omega}{\gamma} v = \dfrac{\psi}{\gamma} & \text{on } \partial\Omega \end{cases}$$

which exists by Proposition 3.1.3. Then for every $(w, w|_{\partial\Omega}) \in \mathcal{V}$ we obtain that

$$\begin{aligned} &\mathfrak{a}\Big((v, v|_{\partial\Omega}) - (u, u|_{\partial\Omega}), (w, w|_{\partial\Omega})\Big) + \omega\left((v, v|_{\partial\Omega}) - (u, u|_{\partial\Omega}) \mid (w, w|_{\partial\Omega})\right)_{\mathcal{H}} \\ &\quad = \int_\Omega f_0 w + \int_\Omega f_j D_j w + \int_{\partial\Omega} g w, \end{aligned}$$

i.e., weakly we have that

$$\begin{cases} L(v-u) + \omega(v-u) = f_0 - \displaystyle\sum_{j=1}^N D_j f_j & \text{on } \Omega, \\ \dfrac{\partial(v-u)}{\partial\nu_L} + \dfrac{\beta+\omega}{\gamma}(v-u) = g + \displaystyle\sum_{j=1}^N f_j \nu_j & \text{on } \partial\Omega. \end{cases} \tag{4.2.3}$$

Since by Example 3.2.7 the solution operator maps appropriate right-hand-side spaces into $C^{0,\alpha}(\Omega)$, $\alpha \in (0,1)$, and thus in particular into $C(\overline{\Omega})$, the closed graph theorem asserts that the solution operator is bounded between these spaces. Since the right hand side of (4.2.3) is small by construction, we can conclude that

$$\|(u, u|_{\partial\Omega}) - (v, v|_{\partial\Omega})\|_{\mathscr{C}} = \|u - v\|_{L^\infty(\Omega)} \leq c\varepsilon$$

for a constant c that does not depend on ε, u, or v. Since

$$(v, v|_{\partial\Omega}) \in D(A + \omega) = D(A)$$

by definition, letting ε tend to zero we find a family of functions in $D(A^c)$ converging to $(u, u|_{\partial\Omega})$ in the norm of $\mathscr{C}$. Since $\mathscr{C}^\infty$ is dense in $C(\overline{\Omega})$, this shows that A^c is densely defined. □

Theorem 4.2.6. *Problem 4.2.1 is well-posed on* $C(\overline{\Omega})$, *meaning that the operator* $-A^c$ *defined in Lemma 4.2.4 generates a* C_0*-semigroup on the isometric copy* $\mathscr{C}$ *of* $C(\overline{\Omega})$ *in* $\mathcal{H}$. *Thus for* $u_0 \in C(\overline{\Omega})$ *the unique mild solution* $U\colon [0,\infty) \to \mathcal{H}$ *of* (4.2.1) *is of the form* $U(t) = (u(t), u(t)|_{\partial\Omega})$, *where* u *is continuous as a function from* $[0,\infty)$ *to* $C(\overline{\Omega})$. *In other words, we can choose representatives of* $u(t)$, $t \geq 0$, *in such a way that* $(t,x) \mapsto u(t)(x)$ *is continuous on* $[0,\infty) \times \overline{\Omega}$.

Proof. Since by Lemmata 4.2.4 and 4.2.5 the operator $-A^c$ satisfies the assumptions of Theorem 1.7.2, this operator generates a C_0-semigroup $(T^c(t))_{t\geq 0}$ on $\mathscr{C}$. Since semigroups can be expressed in terms of their generators via (1.7.3) and since the resolvents of A^c and A coincide on $\mathscr{C}$, $T^c(t)$ is the restriction of $T(t)$ to $\mathscr{C}$. Thus the mild solution

$$U(t) = T(t)(u_0, u_0|_{\partial\Omega}) = T^c(t)(u_0, u_0|_{\partial\Omega})$$

of (4.2.1) is continuous with values in $\mathscr{C}$, i.e., $U(t) = (u(t), u(t)|_{\partial\Omega})$ with u being continuous with values in $C(\overline{\Omega})$. □

In the comments preceding Lemma 4.2.3 we have seen that one cannot expect the Wentzell-Robin semigroup to be quasi-contractive in $L^\infty(\Omega) \oplus_\infty L^\infty(\partial\Omega)$ unless the coefficients b_j are Lipschitz continuous. On the other hand, it can be shown by direct methods that the semigroup is quasi-contractive in $L^p(\Omega) \oplus_p L^p(\partial\Omega)$ for every $p \in (1,\infty)$ without additional assumptions on b_j, where

$$\|(u,v)\|_{L^p(\Omega)\oplus_p L^p(\partial\Omega)} := \left(\|u\|^p_{L^p(\Omega)} + \|v\|^p_{L^p(\partial\Omega;\frac{d\sigma}{\gamma})}\right)^{1/p}.$$

This result can also be deduced from the corresponding assertion for the Robin semigroup by an elegant argument. It is particularly striking that Theorem 1.7.9 can be applied without having an explicit description of the orthogonal projection that appears in the conditions.

Theorem 4.2.7. *For every $p \in (1,\infty)$ there exists $\omega \in \mathbb{R}$ such that*

$$\left\| T(t)U \right\|_{\mathcal{L}^p} \leq \mathrm{e}^{\omega t} \left\| U \right\|_{\mathcal{L}^p}. \tag{4.2.4}$$

holds for all $U \in \mathcal{L}^p := L^p(\Omega) \oplus_p L^p(\partial\Omega, \frac{\mathrm{d}\sigma}{\gamma})$.

Proof. We consider the spaces that occur here as L^p-spaces over the measure space $(\overline{\Omega}, \mu)$ defined in the proof of Lemma 4.2.3.

Let at first $p \in [2,\infty)$, and let C denote the unit ball in $\mathcal{L}^p$. Then C is a convex, closed subset of $\mathcal{H}$ by Lemma 1.4.2. We denote by P_C the orthogonal projection of $\mathcal{H}$ onto C.

By Theorem 1.6.8, the space $\mathcal{L}^p = L^p(\overline{\Omega}, \mu)$ is a real interpolation space between $\mathcal{H}$ and $L^\infty(\overline{\Omega}, \mu)$. Thus, by Corollary 1.6.11, if (4.2.4) is fulfilled for $p = 2$ and $p = \infty$, then also for every other p in-between. For $p = 2$, estimate (4.2.4) is always satisfied, compare Remark 1.7.8. For $p = \infty$ the claim is satisfied at least if the functions b_j are Lipschitz continuous, see Lemma 4.2.3. Thus there exists a Wentzell-Robin semigroup $(S(t))_{t\geq 0}$ and $\omega_S \in \mathbb{R}$ such that

$$\left\| S(t)U \right\|_{L^p(\Omega)\oplus_p L^p(\partial\Omega)} \leq \mathrm{e}^{\omega_S t} \left\| U \right\|_{L^p(\Omega)\oplus_p L^p(\partial\Omega)}.$$

for all $U \in \mathcal{L}^p$, i.e., such that $(\mathrm{e}^{-\omega_S t} S(t))_{t\geq 0}$ leaves C invariant. Thus it follows from Theorem 1.7.9 that

$$P_C \mathcal{V} \subset \mathcal{V}. \tag{4.2.5}$$

Now let $(T(t))_{t\geq 0}$ be an arbitrary Wentzell-Robin semigroup with associated form $\mathfrak{a}_W$ defined on $\mathcal{V} \times \mathcal{V}$. We write $(R(t))_{t\geq 0}$ for the Robin semigroup that comes from the form

$$\mathfrak{a}_R(u,v) := \mathfrak{a}_W\Big((u, u|_{\partial\Omega}), (v, v|_{\partial\Omega})\Big) \tag{4.2.6}$$

defined on $W^{1,2}(\Omega) \times W^{1,2}(\Omega)$. By a result due to Daners [Dan00a, Theorem 5.1], there exists $\omega \geq 0$ such that $(\mathrm{e}^{-\omega t} R(t))_{t\geq 0}$ leaves the unit ball B of $L^p(\Omega)$ invariant. Thus by Theorem 1.7.9,

$$\mathfrak{a}_R(P_B u, u - P_B u) + \omega\,(P_B u \mid u - P_B u)_{L^2(\Omega)} \geq 0 \text{ for all } u \in W^{1,2}(\Omega), \tag{4.2.7}$$

where P_B denotes the orthogonal projection in $L^2(\Omega)$ onto B. Since obviously

$$P_C(u,v) = (P_B u, P_{B'} v)$$

for all $(u,v) \in \mathcal{H}$ with $P_{B'}$ being the orthogonal projection in $L^2(\partial\Omega, \frac{\mathrm{d}\sigma}{\gamma})$ onto the unit ball B' of $L^p(\partial\Omega, \frac{\mathrm{d}\sigma}{\gamma})$, we can deduce from (4.2.5) that

$$P_C(u, u|_{\partial\Omega}) = (P_B u, (P_B u)|_{\partial\Omega}) \text{ for all } u \in W^{1,2}(\Omega). \tag{4.2.8}$$

Combining (4.2.6), (4.2.7), and (4.2.8), we see that

$$\mathfrak{a}_W(P_C U, U - P_C U) + \omega (P_C U \mid U - P_C U)_{\mathcal{H}} \geq \omega (P_{B'} v \mid v - P_{B'} v)_{L^2(\partial\Omega, \frac{d\sigma}{\gamma})} \geq 0 \quad (4.2.9)$$

for all $U = (u, v) = (u, u|_{\partial\Omega}) \in \mathcal{V}$, the last expression being non-negative since $P_{B'}$ is an orthogonal projection and $0 \in B'$. From (4.2.5) and (4.2.9) it follows by Theorem 1.7.9 that $(\mathrm{e}^{-\omega t} T(t))_{t \geq 0}$ leaves C invariant, which is precisely (4.2.4).

Applying the result for $p \geq 2$ to the adjoint equation, we obtain that (4.2.4) remains true if instead of $p \in [2, \infty)$ we require $p' \in [2, \infty)$, i.e., if $p \in (1, 2]$. □

4.3 Nonlinear equations with Robin boundary conditions

The nonlinear situation is more difficult than the linear one, which is why we restrict ourselves to only two special situations, namely our model cases from the previous chapters. We treat a class of Robin boundary conditions that admit Neumann boundary conditions as a special case. The same techniques would allow to handle similar operators and to admit variable coefficients or (monotone) lower order terms, but this would make the arguments more cumbersome.

On a bounded Lipschitz domain $\Omega \subset \mathbb{R}^N$, we investigate the parabolic equations

$$\begin{cases} \dot{u}(t,x) - \Delta_m u(t,x) = 0, & t > 0, x \in \Omega, \\ |\nabla u(t,z)|^{m-2} \dfrac{\partial u}{\partial \nu}(t,z) + h(u(t,z)) = 0, & t > 0, z \in \partial\Omega, \\ u(0,x) = u_0(x), & x \in \Omega, \end{cases}$$

where $\Delta_m := \operatorname{div}(|\nabla u|^{m-2} \nabla u)$ is the m-Laplace operator, and

$$\begin{cases} \dot{u}(t,x) - \operatorname{div}\Big((1 + \arctan(|\nabla u(t,x)|^2)) \nabla u(t,x)\Big) = 0, & t > 0, x \in \Omega, \\ (1 + \arctan(|\nabla u(t,z)|^2)) \dfrac{\partial u}{\partial \nu}(t,z) + h(u(t,z)) = 0, & t > 0, z \in \partial\Omega, \\ u(0,x) = u_0(x), & x \in \Omega. \end{cases}$$

To be able to work with both problems at once, we notice that both problems are of the form

$$\begin{cases} \dot{u}(t,x) - \operatorname{div}\Big(w(|\nabla u(t,x)|^2) \nabla u(t,x)\Big) = 0, & t > 0, x \in \Omega, \\ w(|\nabla u(t,z)|^2) \dfrac{\partial u}{\partial \nu}(t,z) + h(u(t,z)) = 0, & t > 0, z \in \partial\Omega, \\ u(0,x) = u_0(x), & x \in \Omega \end{cases} \quad (4.3.1)$$

for some function w. In addition, we impose some conditions onto h. For easier reference, we collect the assumptions.

Assumption 4.3.1. *Let $\Omega \subset \mathbb{R}^N$ be a bounded Lipschitz domain. Let $w\colon (0,\infty) \to (0,\infty)$ be either $w(x) = x^{(m-2)/2}$ for some $m \in (1, N)$ or $w(x) = 1 + \arctan(x)$. If in the first case we have $m < 2$, then in addition we assume that (3.2.12) holds, i.e., that m is large enough such that Theorem 3.2.14 applies. In the second case we assume $N \geq 3$ and set $m := 2$. Let $h\colon \mathbb{R} \to \mathbb{R}$ be continuous and non-decreasing. Assume moreover that there exists $c \geq 0$ such that*

$$h(x) \leq c|x|^{m-1} \tag{4.3.2}$$

for all $x \in \mathbb{R}$.

As in the linear case, the goal is to show that (4.3.1) admits a unique solution in a sufficiently weak sense and that this solution is jointly continuous in time and space up to the boundary of $[0,\infty) \times \overline{\Omega}$. In other words, we show that the elliptic operator generates consistent nonlinear contraction C_0-semigroups on $L^2(\Omega)$ and $C(\overline{\Omega})$.

We start by showing that (4.3.1) is governed by a nonlinear contraction C_0-semigroup on $L^2(\Omega)$ whose generator is a subdifferential.

Lemma 4.3.2. *Under Assumption 4.3.1, define*

$$\begin{aligned}
W(x) &:= \int_0^x w(t)\,\mathrm{d}t \text{ for } x \geq 0,\\
H(x) &:= \int_0^x h(t)\,\mathrm{d}t \text{ for } x \in \mathbb{R}, \text{ and}\\
\varphi(u) &:= \begin{cases} \frac{1}{2}\int_\Omega W(|\nabla u|^2) + \int_{\partial\Omega} H(u), & u \in W^{1,m}(\Omega) \cap L^2(\Omega),\\ \infty, & u \in L^2(\Omega) \setminus W^{1,m}(\Omega).\end{cases}
\end{aligned}$$

Then $\varphi\colon L^2(\Omega) \to (-\infty,\infty]$ is proper, convex, and lower semicontinuous. Moreover, for u and f in $L^2(\Omega)$ the following assertions are equivalent:

(i) $u \in D(\partial\varphi)$ and $f \in \partial\varphi(u)$.

(ii) u is a weak solution in $W^{1,m}(\Omega)$ of

$$\begin{cases} -\operatorname{div}(w(|\nabla u|^2)\nabla u) = f & \text{on } \Omega\\ w(|\nabla u|^2)\dfrac{\partial u}{\partial \nu} + h(u) = 0 & \text{on } \partial\Omega. \end{cases} \tag{4.3.3}$$

In particular, the subdifferential $\partial\varphi$ of φ is single-valued.

Proof. Let $V := W^{1,m}(\Omega) \cap L^2(\Omega)$ be equipped with the norm

$$\|u\|_V := \|u\|_{L^2(\Omega)} + \|\nabla u\|_{L^m(\Omega)^N}.$$

Then V is a reflexive Banach space. In fact, V can be regarded as a closed subspace of $L^2(\Omega) \oplus L^m(\Omega)^N$.

We split φ into its two parts φ_1 and φ_2 and study these parts on V. We start, however, with

$$\psi_1(u) := \frac{1}{2}\int_\Omega W(|u|^2)$$

regarded as a functional $\psi_1 \colon L^m(\Omega)^N \to (-\infty, \infty]$. Here we equip $L^m(\Omega)^N$ with the norm

$$\|u\|_{L^m(\Omega)^N} = \Big(\int_\Omega |u|^m\Big)^{1/m},$$

where $|\cdot|$ denotes the Euclidean norm in $\mathbb{R}^N$. Since $\frac{\mathrm{d}}{\mathrm{d}x}\frac{1}{2}W(x^2) = w(x^2)x$ is non-decreasing, $x \mapsto W(x^2)$ and hence ψ_1 is convex. As in the proof of Lemma 1.4.3 one can show that $v \mapsto W(|v|^2)$ is continuous from $L^m(\Omega)$ to $L^1(\Omega)$ and $u \mapsto |u|$ is continuous from $L^m(\Omega)^N$ to $L^m(\Omega)$. Hence $\psi_1 \colon L^m(\Omega)^N \to [0, \infty)$ is continuous.

We show that ψ_1 is Gâteaux-differentiable at every point $u \in L^m(\Omega)^N$ with Gâteaux derivative

$$\langle \psi_1'(u), v\rangle_{L^{m'}(\Omega)^N, L^m(\Omega)^N} = \int_\Omega w(|u|^2)u \cdot v. \tag{4.3.4}$$

for $v \in L^m(\Omega)^N$. To see this, note that by applying the mean value theorem we obtain that

$$\frac{\psi_1(u+tv) - \psi_1(u)}{t} - \int_\Omega w(|u|^2)u \cdot v = \int_\Omega \Big(w(|u+\xi v|^2)(u+\xi v)\cdot v - w(|u|^2)u \cdot v\Big)$$

with $\xi \in (0,t)$ depending on u, v, t, and the integration variable. The integrand of the right hand side tends to zero pointwise as $t \to 0$. Since for both choices of w that we allowed in Assumption 4.3.1 we can majorize the integrand, Lebesgue's dominated convergence theorem yields that the integral tends to zero as $t \to 0$. We have thus shown (4.3.4).

Now consider the bounded linear operator $D \colon V \to L^m(\Omega)^N$, $v \mapsto \nabla v$. Define $\varphi_1 := \psi_1 \circ D$. Then $\varphi_1 \colon V \to [0, \infty)$ is proper, convex, continuous, and Gâteaux differentiable. More precisely, the Gâteaux derivative of φ_1 is given by

$$\langle \varphi_1'(u), v\rangle_{V',V} = \langle \psi_1'(Du), Dv\rangle_{L^{m'}(\Omega)^N, L^m(\Omega)^N} = \int_\Omega w(|\nabla u|^2)\nabla u \cdot \nabla v$$

for u and v in V.

Now define

$$\psi_2(u) := \int_{\partial\Omega} H(u)$$

for $u \in L^m(\partial\Omega)$. Note that this expression is well-defined since from (4.3.2) we obtain that $|H(x)| \le \frac{c}{m}|x|^m$ for all $x \in \mathbb{R}$. Since H is convex and continuous, the functional ψ_2 is convex and continuous, see again the proof of Lemma 1.4.3. Moreover, again applying Lebesgue's dominated convergence theorem with a majorizing function obtained from (4.3.2) we see that ψ_2 is Gâteaux differentiable with derivative

$$\langle \psi_2'(u), v\rangle_{L^m(\Omega),L^{m'}(\Omega)} = \int_{\partial\Omega} h(u)v.$$

Let T denote the trace operator $T\colon V \to L^m(\Omega)$, $v \mapsto v|_{\partial\Omega}$. Then $\varphi_2 := \psi_2 \circ T$ is proper, convex, continuous, and Gâteaux differentiable with derivative

$$\langle \varphi_2'(u), v\rangle_{V',V} = \int_{\partial\Omega} h(u)v.$$

Thus $\varphi_1 + \varphi_2$ is proper, convex, continuous, and Gâteaux differentiable on V with derivative

$$\langle (\varphi_1 + \varphi_2)'(u), v\rangle_{V',V} = \int_\Omega w(|\nabla u|^2)\nabla u \cdot \nabla v + \int_{\partial\Omega} h(u)v.$$

Hence by Lemma 1.8.9 the functional φ is proper, convex, and lower semicontinuous, $\partial\varphi$ is single-valued, and $f \in \partial\varphi(u)$ if and only if

$$\int_\Omega w(|\nabla u|^2)\nabla u \nabla v + \int_{\partial\Omega} h(u)v = \int_\Omega fv$$

for all $v \in V$, i.e., since V is dense in $C^\infty(\overline{\Omega})$ if and only if $u \in L^2(\Omega)$ is a weak solution in $W^{1,m}(\Omega)$ of (4.3.3). □

The previous lemma justifies the following definition.

Definition 4.3.3. Under Assumption 4.3.1, let φ be as in Lemma 4.3.2 and define $A := \partial\varphi$. We call a continuous function $u\colon [0,\infty) \to L^2(\Omega)$ a *mild solution of* (4.3.1) *for the initial value* $u_0 \in L^2(\Omega)$ if $u(0) = u_0$ and u is a mild solution of $u'(t) + Au(t) = 0$ in the sense of Definition 1.8.5.

In fact, if u is mild solution of (4.3.1), then by Theorem 1.8.10 we have $u(t) \in D(A)$ for all $t > 0$, implying that $u(t)$ solves (4.3.3) for $f = Au(t) \in L^2(\Omega)$. So $u(t)$ satisfies the Robin boundary conditions in a weak sense.

Since $\mathrm{dom}(\varphi) = W^{1,m}(\Omega) \cap L^2(\Omega)$ is dense in $L^2(\Omega)$, the following is an immediate consequence of Lemma 1.8.8 and Theorem 1.8.10.

Theorem 4.3.4. *Under Assumption 4.3.1, for every $u_0 \in L^2(\Omega)$ there exists a mild solution of* (4.3.1). *Thus* (4.3.1) *is governed by a nonlinear contraction C_0-semigroup $(T(t))_{t\geq 0}$ on $L^2(\Omega)$.*

The next step is to extrapolate the semigroup $(T(t))_{t\geq 0}$ of Theorem 4.3.4 to $L^\infty(\Omega)$.

Lemma 4.3.5. *Let $\lambda > 0$ be arbitrary, and let $H\colon \mathbb{R} \to \mathbb{R}$ be convex. Then*

$$H(\tfrac{1}{2}(x + y + \lambda)) + H(\tfrac{1}{2}(x + y - \lambda)) \leq H(x) + H(y)$$

for all $x, y \in \mathbb{R}$ satisfying $|x - y| \geq \lambda$.

Proof. Since H is convex, there exists a non-decreasing function $h\colon \mathbb{R} \to \mathbb{R}$ such that $H(x) = H(0) + \int_0^x h(t)\,\mathrm{d}t$ for all $x \in \mathbb{R}$, confer for example [RV73, Section 11]. Now let $x, y \in \mathbb{R}$ be such that $|x-y| \geq \lambda$. Without loss of generality we may assume that $y \geq x + \lambda$, interchanging the roles of x and y otherwise. Define $\beta := \frac{1}{2}(y - x - \lambda) \geq 0$. Then

$$\begin{aligned} &H(x) + H(y) - H(\tfrac{1}{2}(x+y+\lambda)) - H(\tfrac{1}{2}(x+y-\lambda)) \\ &\quad = (H(y) - H(y-\beta)) - (H(x+\beta) - H(x)) = \int_{y-\beta}^{y} h(t)\,\mathrm{d}t - \int_{x}^{x+\beta} h(t)\,\mathrm{d}t \\ &\quad = \int_x^{x+\beta} \Big(h(t+y-\beta-x) - h(t)\Big)\mathrm{d}t \geq 0 \end{aligned}$$

since $\beta \leq y - x$, i.e., $t + y - \beta - x \geq t$, and h is non-decreasing. □

Theorem 4.3.6. *Under Assumption 4.3.1, the semigroup* $(T(t))_{t\geq 0}$ *of Theorem 4.3.4 is non-expansive on* $L^\infty(\Omega)$.

Proof. We check the first condition in Theorem 1.8.11 for the functionals

$$\varphi_1(u) := \frac{1}{2}\int_\Omega W(|\nabla u|^2)$$

and

$$\varphi_2(u) := \int_{\partial\Omega} H(u)$$

on $L^2(\Omega)$, where we let $\varphi_1(u) := \infty$ and $\varphi_2(u) := \infty$ for $u \notin W^{1,m}(\Omega)$. Here W and H are defined as in Lemma 4.3.2.

Let $\lambda > 0$. Since the operator P_λ in Theorem 1.8.11 is in fact of the form

$$P_\lambda(u,v)(x) = p_\lambda(u(x), v(x))$$

with a Lipschitz continuous mapping $p_\lambda\colon \mathbb{R}^2 \to \mathbb{R}^2$, the space $W^{1,m}(\Omega) \cap L^2(\Omega)$ is invariant under P_λ by Lemma 1.2.6.

Moreover, by (1.2.1) we obtain for $(f,g) = P_\lambda(u,v)$ that

$$\nabla f = \nabla u\, \mathbb{1}_{\{|u-v|\leq\lambda\}} + \frac{1}{2}(\nabla u + \nabla v)\mathbb{1}_{\{|u-v|>\lambda\}} \tag{4.3.5}$$

and

$$\nabla g = \nabla v\, \mathbb{1}_{\{|u-v|\leq\lambda\}} + \frac{1}{2}(\nabla u + \nabla v)\mathbb{1}_{\{|u-v|>\lambda\}}. \tag{4.3.6}$$

Since $x \mapsto W(x^2)$, where W is defined as in Lemma 4.3.2, is convex and increasing, we have

$$\int_{\{|u-v|>\lambda\}} W\big(|\tfrac{1}{2}\nabla u + \tfrac{1}{2}\nabla v|^2\big) \le \int_{\{|u-v|>\lambda\}} W\big((\tfrac{1}{2}|\nabla u| + \tfrac{1}{2}|\nabla v|)^2\big)$$
$$\le \frac{1}{2}\int_{\{|u-v|>\lambda\}} W(|\nabla u|^2) + \frac{1}{2}\int_{\{|u-v|>\lambda\}} W(|\nabla v|^2).$$

Combining the latter estimate with (4.3.5) and (4.3.6) we see that

$$\varphi_1^{(2)}(f,g) \le \varphi_1^{(2)}(u,v). \tag{4.3.7}$$

It is a direct consequence of Lemma 4.3.5 that

$$\int_{\partial\Omega\cap\{|u-v|>\lambda\}} \Big(H(\tfrac{1}{2}(u+v+\lambda)) + H(\tfrac{1}{2}(u+v-\lambda))\Big) \le \int_{\partial\Omega\cap\{|u-v|>\lambda\}} \Big(H(u)+H(v)\Big).$$

Hence

$$\varphi_2^{(2)}(f,g) \le \varphi_2^{(2)}(u,v). \tag{4.3.8}$$

It follows from (4.3.7) and (4.3.8) that the first assertion of Theorem 1.8.11 is satisfied for the functional φ of Lemma 4.3.2. This proves that $(T(t))_{t\ge 0}$ is non-expansive on $L^\infty(\Omega)$. □

It is easy to deduce from Theorem 4.3.6 that the generator is m-accretive on $\mathrm{C}(\overline{\Omega})$.

Theorem 4.3.7. *Under Assumption 4.3.1, let $(T(t))_{t\ge 0}$ be the semigroup of Theorem 4.3.4 and A be its generator. Then the part A^c of A in $\mathrm{C}(\overline{\Omega})$, i.e.,*

$$D(A^c) := \{u \in \mathrm{C}(\overline{\Omega}) : Au \in \mathrm{C}(\overline{\Omega})\}, \quad A^c u := Au,$$

is m-accretive on $\mathrm{C}(\overline{\Omega})$.

Proof. It follows from Theorems 4.3.6 and 1.8.11 that

$$\|u_1 - u_2\|_{L^\infty(\Omega)} \le \|(u_1 + \alpha A u_1) - (u_2 + \alpha A u_2)\|_{L^\infty(\Omega)}$$

for all $\alpha > 0$ and all $u_1, u_2 \in D(A) \supset D(A^c)$. Thus A^c is accretive on $\mathrm{C}(\overline{\Omega})$.

It remains to show that $I + \alpha A^c \colon \mathrm{C}(\overline{\Omega}) \to \mathrm{C}(\overline{\Omega})$ is surjective for some $\alpha > 0$. Since A is m-accretive on $L^2(\Omega)$, for every $f \in \mathrm{C}(\overline{\Omega}) \subset L^2(\Omega)$ there exists $u \in D(A)$ such that $Au = f$. By Lemma 4.3.2 this implies that u is a weak solution of (4.3.3), hence $u \in \mathrm{C}(\overline{\Omega})$ by Example 3.2.9, 3.2.10, or 3.2.16, respectively. Thus $u \in D(A^c)$ and $A^c u = f$, by which we have proved that A^c is m-accretive on $\mathrm{C}(\overline{\Omega})$. □

Now we know from Theorems 4.3.7 and 1.8.6 that A^c generates a semigroup on $\overline{D(A^c)}$. We show that in fact $\overline{D(A^c)} = \mathrm{C}(\overline{\Omega})$.

Lemma 4.3.8. *Under Assumption 4.3.1, the operator A^c of Theorem 4.3.7 is densely defined on* $\mathrm{C}(\overline{\Omega})$.

Proof. For $\omega \geq 0$, define $A_\omega \colon V \to V'$, $V := W^{1,m}(\Omega) \cap L^2(\Omega)$, by

$$\langle A_\omega u, v\rangle := \int_\Omega w(|\nabla u|^2)\nabla u \nabla v + \omega \int_\Omega |u|^{m-2} uv + \int_{\partial\Omega} h(u)v \tag{4.3.9}$$

for $v \in V$. Then the part A^c_ω of A_ω in $\mathrm{C}(\overline{\Omega})$ has domain $D(A^c_\omega) = D(A^c)$. Thus it suffices to show that there exists $\omega \geq 0$ such that A^c_ω is densely defined.

First assume $h = 0$, i.e., Neumann boundary conditions. Let $\omega > 0$. If $q \in (1, \infty)$ is large enough, then for $f_0 \in L^q(\Omega)$ and $f_j \in L^q(\Omega)$ there exists a unique solution $u \in V$ of

$$\begin{cases} -\operatorname{div}\Big(w(|\nabla u|^2)\nabla u\Big) + \omega |u|^{m-2} u = f_0 - \sum_{j=1}^N D_j f_j & \text{on } \Omega, \\ w(|\nabla u|^2)\dfrac{\partial u}{\partial \nu} = \sum_{j=1}^N f_j \nu_j & \text{on } \partial\Omega, \end{cases} \tag{4.3.10}$$

see Examples 2.2.9 and 2.2.10. Moreover, if q is large enough, then even $u \in \mathrm{C}^{0,\alpha}(\Omega)$ and the Hölder norm of u can be bounded in terms of the norms of f_0 and f_j in $L^q(\Omega)$ by Corollary 2.1.6. Moreover, u depends continuously in $W^{1,m}(\Omega)$ on f_0 and f_j in $L^q(\Omega)$ as seen in Examples 2.3.7, 2.3.20, and 2.3.23. Now the same argument as in Corollary 2.3.19 shows that u depends continuously in $\mathrm{C}(\overline{\Omega})$ on f_0 and f_j in $L^q(\Omega)$.

In the general situation, i.e., $h \neq 0$, essentially the same reasoning goes through. In fact, existence of unique solutions for $\omega > 0$ is shown as in Theorem 3.1.1. Continuous dependence in $W^{1,m}(\Omega)$ follows by the same arguments as for Neumann boundary conditions since the boundary term involving h again induces a monotone operator by monotonicity of h. Finally, the bound in $\mathrm{C}^{0,\alpha}(\Omega)$ can be carried through the bootstrapping argument in Theorem 3.2.5. Thus also in this situation we have continuous dependence in $\mathrm{C}(\overline{\Omega})$ as the functions on the right hand side vary in $L^q(\Omega)$.

Now let $u \in \mathrm{C}^\infty(\overline{\Omega})$ be given. Then $A_\omega u \in W^{1,m}(\Omega)'$. The functions $w(|\nabla u|^2)\nabla u$ and $h(u)$ are bounded, hence $A_\omega u \in W^{1,1}(\Omega)'$. Thus, as in Theorem 1.4.4, there exist f_0 and f_j in $L^\infty(\Omega)$ such that

$$\langle A_\omega u, v\rangle = \int_\Omega f_0\, v + \sum_{j=1}^N \int_\Omega f_j\, D_j v$$

for all $v \in W^{1,m}(\Omega)$, i.e., u solves (4.3.10) for this right hand side. Pick functions f_0^n and f_j^n in $\mathrm{C}^\infty(\overline{\Omega})$ such that $f_0^n \to f_0$ and $f_j^n \to f_j$ in $L^q(\Omega)$ for some $q \in (q_0, \infty)$. Let u^n denote the weak solution in $W^{1,m}(\Omega)$ of (4.3.10) for these right hand sides. Then $u^n \in D(A^c_\omega)$. Moreover, by continuity of the solution operator, $u^n \to u$ in $\mathrm{C}(\overline{\Omega})$, hence $u \in \overline{D(A^c_\omega)}$.

We have shown that $\mathrm{C}^\infty(\overline{\Omega}) \subset \overline{D(A^c_\omega)}$. Since $\mathrm{C}^\infty(\overline{\Omega})$ is dense in $\mathrm{C}(\overline{\Omega})$ by the Stone-Weierstrass theorem, this implies that $D(A^c_\omega) = D(A^c)$ is dense in $\mathrm{C}(\overline{\Omega})$. □

The final step is to show that the semigroup generated by A^c is the restriction of $(T(t))_{t\geq 0}$ to $\mathrm{C}(\overline{\Omega})$. However, this is trivial since every mild solution of $u'(t)+A^c u(t)=0$ on $\mathrm{C}(\overline{\Omega})$ is a mild solution of $u'(t)+Au(t)=0$ on $L^2(\Omega)$, see for example [Sho97, Proposition IV.8.1]. We have proved the following theorem.

Theorem 4.3.9. *Under Assumption 4.3.1, the semigroup $(T(t))_{t\geq 0}$ on $L^2(\Omega)$ with generator A that we introduced in Theorem 4.3.4 restricts to a nonlinear contraction C_0-semigroup on $\mathrm{C}(\overline{\Omega})$ in the sense of Definition 1.8.1. Its generator in the sense of Definition 1.8.5 is the part of A in $\mathrm{C}(\overline{\Omega})$. Thus for $u_0 \in \mathrm{C}(\overline{\Omega})$ the unique mild solution of (4.3.1) is in $\mathrm{C}([0,\infty);\mathrm{C}(\overline{\Omega}))$, thus jointly continuous in time and space.*

Remark 4.3.10. An analogous result is true for the parabolic problem with Wentzell-Robin boundary conditions. In fact, the transition from Robin to Wentzell-Robin boundary conditions can be carried out in the same manner as in Section 4.2.

Bibliography

[ABHN01] W. Arendt, C. J. K. Batty, M. Hieber, and F. Neubrander, *Vector-valued Laplace transforms and Cauchy problems*, Monographs in Mathematics, vol. 96, Birkhäuser Verlag, Basel, 2001.

[AC84] C. Atkinson and C. R. Champion, *Some boundary-value problems for the equation* $\nabla \cdot (| \nabla\varphi |^N \nabla\varphi) = 0$, Quart. J. Mech. Appl. Math. **37** (1984), no. 3, 401–419.

[Agr08] M. S. Agranovich, *Spectral problems in Lipschitz domains for strongly elliptic systems in the Banach spaces* H_p^σ *and* B_p^σ, Funct. Anal. Appl. **42** (2008), no. 4, 249–267.

[AMPR03] W. Arendt, G. Metafune, D. Pallara, and S. Romanelli, *The Laplacian with Wentzell-Robin boundary conditions on spaces of continuous functions*, Semigroup Forum **67** (2003), no. 2, 247–261.

[AR97] J.-J. Alibert and J.-P. Raymond, *Boundary control of semilinear elliptic equations with discontinuous leading coefficients and unbounded controls*, Numer. Funct. Anal. Optim. **18** (1997), no. 3-4, 235–250.

[AtE97] W. Arendt and A. F. M. ter Elst, *Gaussian estimates for second order elliptic operators with boundary conditions*, J. Operator Theory **38** (1997), no. 1, 87–130.

[AW03] W. Arendt and M. Warma, *The Laplacian with Robin boundary conditions on arbitrary domains*, Potential Anal. **19** (2003), no. 4, 341–363.

[BCP] P. Bénilan, M. G. Crandall, and A. Pazy, *Nonlinear evolution equations in banach spaces*, Preprint book.

[Ber84] J. Bergh, *A nonlinear complex interpolation result*, Interpolation spaces and allied topics in analysis (Lund, 1983), Lecture Notes in Math., vol. 1070, Springer, Berlin, 1984, pp. 45–47.

[BH91] R. F. Bass and P. Hsu, *Some potential theory for reflecting Brownian motion in Hölder and Lipschitz domains*, Ann. Probab. **19** (1991), no. 2, 486–508.

[Bie10] M. Biegert, *A priori estimates for the difference of solutions to quasi-linear elliptic equations*, submitted, 2010.

[Böh91] M. Böhm, *Remarks on complex interpolation of some nonlinear operators. I*, Math. Nachr. **153** (1991), 191–206.

[Böh93] ———, *On complex interpolation of nonlinear operators. II*, Boll. Un. Mat. Ital. B (7) **7** (1993), no. 1, 127–148.

[Bré68] H. Brézis, *Équations et inéquations non linéaires dans les espaces vectoriels en dualité*, Ann. Inst. Fourier (Grenoble) **18** (1968), no. fasc. 1, 115–175.

[Bré73] ———, *Opérateurs maximaux monotones et semi-groupes de contractions dans les espaces de Hilbert*, North-Holland Publishing Co., Amsterdam, 1973, North-Holland Mathematics Studies, No. 5. Notas de Matemática (50).

[BW09] M. Biegert and M. Warma, *Some quasi-linear elliptic equations with inhomogeneous generalized Robin boundary conditions on bad domains*, Ulmer Seminare **14** (2009), 43–63.

[BZ88] Je. I. Berezhnoj and P. P. Zabrejko, *Some interpolation theory for nonlinear operators*, Nonlinear Anal. **12** (1988), no. 2, 155–170.

[CFG+08] G. M. Coclite, A. Favini, G. R. Goldstein, J. A. Goldstein, and S. Romanelli, *Continuous dependence on the boundary conditions for the Wentzell Laplacian*, Semigroup Forum **77** (2008), no. 1, 101–108.

[CG03] F. Cipriani and G. Grillo, *Nonlinear Markov semigroups, nonlinear Dirichlet forms and applications to minimal surfaces*, J. Reine Angew. Math. **562** (2003), 201–235.

[Che98] Y. Cheng, *Hölder continuity of the inverse of p-Laplacian*, J. Math. Anal. Appl. **221** (1998), no. 2, 734–748.

[Cwi87] M. Cwikel, *Complex interpolation of nonlinear operators and Lebedev-Milin inequalities for Taylor coefficients of composed analytic functions*, Math. Nachr. **132** (1987), 153–161.

[Dan00a] D. Daners, *Heat kernel estimates for operators with boundary conditions*, Math. Nachr. **217** (2000), 13–41.

[Dan00b] ———, *Robin boundary value problems on arbitrary domains*, Trans. Amer. Math. Soc. **352** (2000), no. 9, 4207–4236.

[Dan09] ———, *Inverse positivity for general Robin problems on Lipschitz domains*, Arch. Math. (Basel) **92** (2009), no. 1, 57–69.

[DD09] D. Daners and P. Drábek, *A priori estimates for a class of quasi-linear elliptic equations*, Trans. Amer. Math. Soc. **361** (2009), no. 12, 6475–6500.

[DG57] E. De Giorgi, *Sulla differenziabilità e l'analiticità delle estremali degli integrali multipli regolari*, Mem. Accad. Sci. Torino. Cl. Sci. Fis. Mat. Nat. (3) **3** (1957), 25–43.

[DiB93] E. DiBenedetto, *Degenerate parabolic equations*, Universitext, Springer-Verlag, New York, 1993.

[DKM07] F. Duzaar, J. Kristensen, and G. Mingione, *The existence of regular boundary points for non-linear elliptic systems*, J. Reine Angew. Math. **602** (2007), 17–58.

[DM09] F. Duzaar and G. Mingione, *Partial differential equations—gradient estimates in non-linear potential theory*, Atti Accad. Naz. Lincei Cl. Sci. Fis. Mat. Natur. Rend. Lincei (9) Mat. Appl. **20** (2009), no. 2, 179–190.

[EG92] L. C. Evans and R. F. Gariepy, *Measure theory and fine properties of functions*, Studies in Advanced Mathematics, CRC Press, Boca Raton, FL, 1992.

[EN00] K.-J. Engel and R. Nagel, *One-parameter semigroups for linear evolution equations*, Graduate Texts in Mathematics, vol. 194, Springer-Verlag, New York, 2000, With contributions by S. Brendle, M. Campiti, T. Hahn, G. Metafune, G. Nickel, D. Pallara, C. Perazzoli, A. Rhandi, S. Romanelli and R. Schnaubelt.

[Eng03] K.-J. Engel, *The Laplacian on $C(\overline{\Omega})$ with generalized Wentzell boundary conditions*, Arch. Math. (Basel) **81** (2003), no. 5, 548–558.

[Fed69] H. Federer, *Geometric measure theory*, Die Grundlehren der mathematischen Wissenschaften, Band 153, Springer-Verlag New York Inc., New York, 1969.

[FGGR02] A. Favini, G. R. Goldstein, J. A. Goldstein, and S. Romanelli, *The heat equation with generalized Wentzell boundary condition*, J. Evol. Equ. **2** (2002), no. 1, 1–19.

[FGGR08] ______, *Wentzell boundary conditions in the nonsymmetric case*, Math. Model. Nat. Phenom. **3** (2008), no. 7, 143–147.

[FT95] M. Fukushima and M. Tomisaki, *Reflecting diffusions on Lipschitz domains with cusps – analytic construction and Skorohod representation*, Potential Anal. **4** (1995), no. 4, 377–408.

[Gol35] M. Golomb, *Zur Theorie der nichtlinearen Integralgleichungen, Integralgleichungssysteme und allgemeinen Funktionalgleichungen*, Math. Z. **39** (1935), no. 1, 45–75.

[GR01] J. A. Griepentrog and L. Recke, *Linear elliptic boundary value problems with non-smooth data: normal solvability on Sobolev-Campanato spaces*, Math. Nachr. **225** (2001), 39–74.

[Gri85] P. Grisvard, *Elliptic problems in nonsmooth domains*, Monographs and Studies in Mathematics, vol. 24, Pitman (Advanced Publishing Program), Boston, MA, 1985.

[GT01] D. Gilbarg and N. S. Trudinger, *Elliptic partial differential equations of second order*, Classics in Mathematics, Springer-Verlag, Berlin, 2001, Reprint of the 1998 edition.

[GZ77] R. Gariepy and W. P. Ziemer, *A regularity condition at the boundary for solutions of quasilinear elliptic equations*, Arch. Rational Mech. Anal. **67** (1977), no. 1, 25–39.

[HDR09] R. Haller-Dintelmann and J. Rehberg, *Maximal parabolic regularity for divergence operators including mixed boundary conditions*, J. Differential Equations **247** (2009), no. 5, 1354–1396.

[Kač60] R. Kačurovski, *On monotone operators and convex functionals*, Uspekhi Mat. Nauk **15** (1960), 213–215.

[KM94] T. Kilpeläinen and J. Malý, *The Wiener test and potential estimates for quasilinear elliptic equations*, Acta Math. **172** (1994), no. 1, 137–161.

[KM08] J. Kristensen and G. Mingione, *Boundary regularity of minima*, Atti Accad. Naz. Lincei Cl. Sci. Fis. Mat. Natur. Rend. Lincei (9) Mat. Appl. **19** (2008), no. 4, 265–277.

[KR09] C. E. Kenig and D. J. Rule, *The regularity and Neumann problem for non-symmetric elliptic operators*, Trans. Amer. Math. Soc. **361** (2009), no. 1, 125–160.

[Lax02] P. D. Lax, *Functional analysis*, Pure and Applied Mathematics (New York), Wiley-Interscience [John Wiley & Sons], New York, 2002.

[Lê07] A. Lê, *On the local Hölder continuity of the inverse of the p-Laplace operator*, Proc. Amer. Math. Soc. **135** (2007), no. 11, 3553–3560 (electronic).

[Lie83] G. M. Lieberman, *The conormal derivative problem for elliptic equations of variational type*, J. Differential Equations **49** (1983), no. 2, 218–257.

[Lie91] ———, *The natural generalization of the natural conditions of Ladyzhenskaya and Ural'tseva for elliptic equations*, Comm. Partial Differential Equations **16** (1991), 311–361.

[Lie92] ———, *The conormal derivative problem for equations of variational type in nonsmooth domains*, Trans. Amer. Math. Soc. **330** (1992), no. 1, 41–67.

[Lin90] P. Lindqvist, *On the equation* $\operatorname{div}(|\nabla u|^{p-2}\nabla u) + \lambda|u|^{p-2}u = 0$, Proc. Amer. Math. Soc. **109** (1990), no. 1, 157–164.

[LT90] P. Le Tallec, *Numerical analysis of viscoelastic problems*, Recherches en Mathématiques Appliquées [Research in Applied Mathematics], vol. 15, Masson, Paris, 1990.

[LU68] O. A. Ladyzhenskaya and N. N. Ural'tseva, *Linear and quasilinear elliptic equations*, Translated from the Russian by Scripta Technica, Inc. Translation editor: Leon Ehrenpreis, Academic Press, New York, 1968.

[Maz70] V. G. Maz'ya, *The continuity at a boundary point of the solutions of quasilinear elliptic equations*, Vestnik Leningrad. Univ. **25** (1970), no. 13, 42–55.

[Maz85] Vladimir G. Maz'ja, *Sobolev spaces*, Springer Series in Soviet Mathematics, Springer-Verlag, Berlin, 1985, Translated from the Russian by T. O. Shaposhnikova.

[Min06] G. Mingione, *Regularity of minima: an invitation to the dark side of the calculus of variations*, Appl. Math. **51** (2006), no. 4, 355–426.

[Min07] ———, *The Calderón-Zygmund theory for elliptic problems with measure data*, Ann. Sc. Norm. Super. Pisa Cl. Sci. (5) **6** (2007), no. 2, 195–261.

[Min10] ———, *Gradient estimates below the duality exponent*, Math. Ann. **346** (2010), no. 3, 571–627.

[MM09] M. Mitrea and S. Monniaux, *On the analyticity of the semigroup generated by the Stokes operator with Neumann-type boundary conditions on Lipschitz subdomains of Riemannian manifolds*, Trans. Amer. Math. Soc. **361** (2009), no. 6, 3125–3157.

[Mos60] J. Moser, *A new proof of De Giorgi's theorem concerning the regularity problem for elliptic differential equations*, Comm. Pure Appl. Math. **13** (1960), 457–468.

[MT09] M. Mitrea and M. Taylor, *Local regularity results for second order elliptic systems on Lipschitz domains*, Funct. Approx. Comment. Math. **40** (2009), 175–184.

[Nas58] J. Nash, *Continuity of solutions of parabolic and elliptic equations*, Amer. J. Math. **80** (1958), 931–954.

[Nit09a] R. Nittka, *Approximation of the semigroup generated by the Robin Laplacian in terms of the Gaussian semigroup*, J. Funct. Anal. **257** (2009), no. 5, 1429–1444.

[Nit09b] ———, *Regularity of Solutions of Linear Second Order Elliptic and Parabolic Boundary Value Problems on Lipschitz Domains*, submitted, 2009.

[Ouh05] E. M. Ouhabaz, *Analysis of Heat Equations on Domains*, London Mathematical Society Monographs Series, vol. 31, Princeton University Press, 2005.

[Pee70] J. Peetre, *Interpolation of Lipschitz operators and metric spaces*, Mathematica (Cluj) **12(35)** (1970), 325–334.

[Phi61] J. R. Philip, *n-diffusion*, Austral. J. Phys. **14** (1961), 1–13.

[RV73] A. W. Roberts and D. E. Varberg, *Convex functions*, Pure and Applied Mathematics, vol. 57, Academic Press [A subsidiary of Harcourt Brace Jovanovich, Publishers], New York-London, 1973.

[Ser64] J. Serrin, *Local behavior of solutions of quasi-linear equations*, Acta Math. **111** (1964), 247–302.

[She08] Z. Shen, *Recent progress on the Dirichlet problem in Lipschitz domains*, Third International Congress of Chinese Mathematicians. Part 1, 2, AMS/IP Stud. Adv. Math., 42, pt. 1, vol. 2, Amer. Math. Soc., Providence, RI, 2008, pp. 483–491.

[Sho97] R. E. Showalter, *Monotone operators in Banach space and nonlinear partial differential equations*, Mathematical Surveys and Monographs, vol. 49, American Mathematical Society, Providence, RI, 1997.

[Ste70] E. M. Stein, *Singular integrals and differentiability properties of functions*, Princeton Mathematical Series, No. 30, Princeton University Press, Princeton, N.J., 1970.

[Tar72] L. Tartar, *Interpolation non linéaire et régularité*, J. Functional Analysis **9** (1972), 469–489.

[Tol83] P. Tolksdorf, *On the Dirichlet problem for quasilinear equations in domains with conical boundary points*, Comm. Partial Differential Equations **8** (1983), no. 7, 773–817.

[Tri95] H. Triebel, *Interpolation theory, function spaces, differential operators*, second ed., Johann Ambrosius Barth, Heidelberg, 1995.

[Tro87] G. M. Troianiello, *Elliptic differential equations and obstacle problems*, The University Series in Mathematics, Plenum Press, New York, 1987.

[Tru67] N. S. Trudinger, *On Harnack type inequalities and their application to quasilinear elliptic equations*, Comm. Pure Appl. Math. **20** (1967), 721–747.

[War06] M. Warma, *The Robin and Wentzell-Robin Laplacians on Lipschitz domains*, Semigroup Forum **73** (2006), no. 1, 10–30.

[Wer00] D. Werner, *Funktionalanalysis*, extended ed., Springer-Verlag, Berlin, 2000.

[Woo07] I. Wood, *Maximal L^p-regularity for the Laplacian on Lipschitz domains*, Math. Z. **255** (2007), no. 4, 855–875.

[Zar60] E. H. Zarantonello, *Solving functional equations by contractive averaging*, Tech. Report 160, Mathematical Research Center, Madison, Wisconsin, 1960.

[Zie89] W. P. Ziemer, *Weakly differentiable functions*, Graduate Texts in Mathematics, vol. 120, Springer-Verlag, New York, 1989.

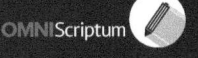

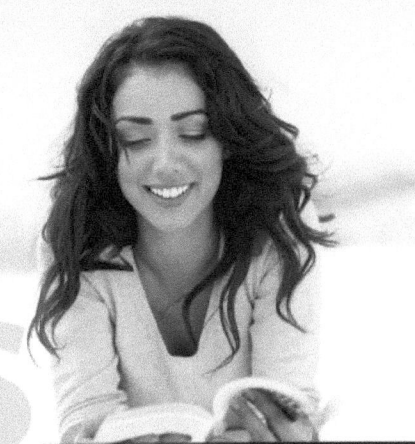

MIX
Papier aus verantwortungsvollen Quellen
Paper from responsible sources
FSC® C105338

Printed by Books on Demand GmbH, Norderstedt / Germany